COMMUNITIES AND FOREST MANAGEMENT IN WESTERN EUROPE

Sally Jeanrenaud

A REGIONAL PROFILE OF

WG-CIFM

THE WORKING GROUP ON COMMUNITY

INVOLVEMENT IN FOREST MANAGEMENT

In memory of Tim Stead (1952-2000) whose creative energy inspired and informed community forestry in the Borders of Scotland and beyond.

FOREWORD TO THE REGIONAL PROFILE SERIES

This series of regional assessments was initiated by an international group of individuals concerned about the future of the world's forests. We began meeting during the sessions of the Intergovernmental Panel of Forests (IPF) convened by the United Nations in New York and Geneva between 1996 and 1997. In order to promote regional exchange and better inform international policy dialogues, we formed the Working Group on Community Involvement in Forest Management (WG-CIFM). The World Conservation Union (IUCN) agreed to facilitate our activities and administer financial support for the Working Group's "Seeking Connections" project, which was provided by the Ford Foundation and the United Kingdom's Department for International Development (DFID).

The Working Group currently includes forest administrators, planning officers, forest scientists, environmental activists, and diplomats. Our discussions of the underlying causes of deforestation and promising strategies to bring greater stability to the world's forests revealed many similarities between our regions. Most group members agreed that the expansion of government and private industry control over forests in the past century had increasingly undermined the management role of communities in their nations. In some cases this was reflected in the deterioration of indigenous forms of resource stewardship, in others policies did not allow for localised systems of forest rights and responsibilities to be established. Many participants reported that a growing number of communities in their countries are attempting to gain greater control over their forest resources. Nations in both the South and the North are beginning to address this imbalance by developing policies and programs to re-engage communities in forest management decision-making.

During the meetings of the Working Group we noted that many government forestry agencies are under-financed, their budgets cut over the past decade due to political changes and economic restructuring in both developed and developing countries. While the rapidly shrinking public forest base is under unprecedented pressure from industry as well from local and urban public forest consumers, many forestry agencies have been faced with severe financial constraints and staff reductions that frustrated their attempts to sustainably manage their national forests. Economic recessions and government downsizing have been catalysts for innovative solutions to forest management problems.

Working side by side with local communities, some forest agencies are forging new partnerships and approaches to forest management. While the subtle pace of this change cannot stem the criticism from conservationists, industry, and local communities not experiencing change, the dialogues and partnerships have sparked a new dynamic animated by citizen's coalitions and regional processes incorporating diverse stakeholder groups. Our group concluded that these parallels warranted a sharing of community forest management experiences between countries in the hope of accelerating the development of more effective strategies to engage forest stakeholders in sustainable forest management.

Throughout the process of the International Panel on Forests, the Working Group sought to introduce language to the draft recommendations that could contribute toward creating new policies that support greater community involvement in forest management. The Working Group convened six times during the meetings of the Intergovernmental Panel on Forests between 1996 and 1998. Over 150 individuals have participated in these sessions. The Working Group was able to effectively influence the final text of the IPF that resulted in some 135 proposals for action approved by governments in June 1997 at the United Nations General Assembly Special Session.

In order to extend our exchange to colleagues and other interested readers who were unable to participate in the Working Group, we decided to establish a monograph series that characterises some of the diverse community forest management experiences from each of the world's regions, emphasising community perspectives. We defined "community" broadly to include small forest-dependent settlements, indigenous peoples, as well as the greater civil society. This broad definition presented the challenge of capturing the inevitable diversity of opinion present in the realm of forest "stakeholders", literally, all those who have ties to or needs that are met through forest environments. Members of the Working Group agreed that the profiles should reflect a range of views of communities, planners, foresters and other stakeholders within each country. The profiles attempt to be both a synthesis and a mosaic of these complex and diverse national and regional realities.

The degree to which community involvement in forest management (CIFM) is recognised by governments and is integrated into state management goals varies widely. Presently, much of the world's forests are used by local communities, whose interactions are mediated through institutions that range from highly traditional to very modern, and whose legal control ranges from nothing to absolute. Because community forest management is often based on local organisations that are frequently unregistered and fall outside policies and prescriptions, local forest-dependent inhabitants have been the hidden component of management in the forestry sector. The communities' role may extend from passive engagement to active participation in goal identification, objective setting, controlling implementation, and assessing results. In some areas community involvement and authority may be comprehensive, based on granted legal autonomy or simple isolation. In 1997, the Working Group developed the following chart to reflect the broad spectrum of ways in which communities interface with government management strategies and the varying levels of authority they may hold.

The goal of the regional profile series is to communicate CIFM experiences between regions, targeting diverse audiences including international policy makers and national planners who are responsible for shaping forest management policies and strategies, as well as the forestry practitioners and development specialists who implement them. To familiarise our cross-cultural audience with the national

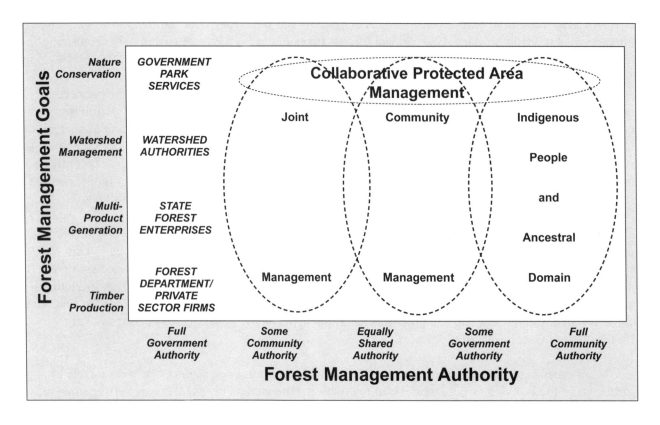

contexts, each regional profile provides a brief summary of the region's forest management history, human ecology, and administrative organisations, followed by a series of CIFM case studies.

Each regional profile is compiled with the collaboration of many individuals and organisations engaged in the countries of the region under review. The contributors include a mix of generalists and in-country specialists who draw on an extensive collection of existing histories, policy reviews, ecological assessments, personal interviews, and case materials. During the assimilation of materials for review, the editor and the contributors participate in national and regional meetings to capture contemporary views and policy trends. Outside reviewers read and comment on a succession of draft manuscripts to better ensure a balanced presentation. Nonetheless, given the controversial nature of the forest policy debate, numerous differences over the interpretation of data or the validity of information are likely to occur. For this reason, the Working Group feels that it is important to act independently of any organisation or institution. I hope our readers find these materials useful in seeking new solutions to forest management issues.

—Simon Rietbergen, Project Co-ordinator

SUMMARY

This profile explores the diverse and changing nature of community involvement in forest management in Western Europe. It provides opportunities for examining the adaptations of CIFM to industrial contexts, as well as envisioning the contributions of CIFM to sustainable futures in Europe's emerging post-industrial context.

The history of people-forest relations in Europe reveals trends familiar to many regions. While forests have provided essential resources for many of the major transformations in social history, it has usually been at a cost to natural ecosystems and to poorer rural communities. Many traditional community based institutions were undermined as they came into conflict with newly emerging state and private interests, from the 17th century. Most natural forest ecosystems have been degraded by the intensification of agricultural and forestry practices, and urbanisation, leaving only a tiny fraction of semi-natural forests.

The book provides some comparative European-level data on important social institutions which shape patterns of community involvement in forestry: tenure arrangements; policy frameworks; forest governance structures, and economic incentives. It briefly examines different national contexts for CIFM. While there are hopeful signs of a transition in policies and activities, analysis suggests significant historical and institutional constraints and powerful resistance to greater CIFM by some groups.

The main part of the profile contains 12 case studies covering various aspects of CIFM in the region:

- The Forests of the Val di Fiemme in Northern Italy, where traditional institutions support thriving communities.

- The Baldios in Portugal, where communities have reclaimed the forested commons.

- Swedish Common Forests, showing how rural communities have adapted to urban, industrialised society.

- Crofter Forestry in Scotland, which has established the rights to manage woodlands.

- Saami Reindeer Herders, who are losing traditional grazing rights in Swedish forests.

- Community Forestry in the Borders Region of Scotland, which is contributing to native woodland restoration.

- Communities and Biodiversity Conservation in the Mediterranean Region, which indicates the under-utilised potential of NTFPs.

- An Urban Community Forestry in London, which emphasises the important role small scale wood processing plays in rejuvenating woodland management.

- Forest Workers in Europe and the role of Unions in decision making.

- Small Forest Owners in Europe and the role of associations in forest management.

- French Forest Communes with opportunities and constraints for community involvement.

- Encouraging Involvement in Public Forest Management, with examples from Finland, Denmark and Switzerland

The final chapters discuss some of the main economic, social, ecological and policy opportunities and challenges of CIFM in Europe in the future, in the light of 12 cases studies, and outline the principal lessons learned according to three key groups of actors: governments, NGOs, and local communities. Some examples of CIFM can be seen as significant spearheads of sustainability in post-industrial Europe.

The profile proposes the following recommendations for policy and action in Europe. It supports:

- A diversity of approaches to community involvement in forest management in Europe;
- Policy reform that emphasises the three interlinked goals of sustainability: economic viability; social equity and environmental protection;
- Secure forest property and usufruct rights;
- Participatory approaches to local, co-management and public forest governance;
- Economic reforms for sustainable rural and urban livelihoods;
- Partnerships and coalitions for community involvement in forest management;
- Forest agency reform;
- Intersectoral coherence and integration of land use policies, such as farming, forestry, tourism and sustainable rural businesses.

A vision for the 21st century is to see CIFM in Europe as a way of (re) connecting people with forests in rural and urban areas for sustainable futures, and as a means of integrating economic, social, cultural, spiritual and ecological values in diverse, innovative, and evolving ways.

ACKNOWLEDGEMENTS

The European profile on communities and forests is the product of many shared insights and interpretations, and heart felt thanks go to numerous individuals who contributed to this learning process.

In particular, I am indebted to a core team who met on several occasions to share experiences, help identify key concepts and frame problems and solutions. Their insights, guidance and enthusiasm have provided a steady source of practical help and moral support throughout the development of the profile. These include: Nanna Borchert; Jill Bowling; Andrea Finger; Mandy Haggith; Jean-Paul Jeanrenaud; Yves Kazemi; Simon Rietbergen; Bill Ritchie; Dagmar Timmer; Kadi Warner and Pier Carlo Zingari.

Special thanks go to the contributors of the case study materials and reviewers of the profile: Nanna Borchet and Olaf Johansson (Saami Reindeer Herders and Forests in Sweden); Jill Bowling (Forest Workers); Roland Brouwer (the Baldios in Portugal); Lars Carlsson (Swedish Forest Commons); Bruno Crosignani (Val di Fiemme in Northern Italy); Nigel Dudley; Andrea Finger (French Forest Communes), Mandy Haggith and Bill Ritchie (Crofter Forestry in Scotland); Natalie Hufnagl (Small Forest Owner Associations); Yves Kazemi (Public Participation in Swiss Forests); Christian Küchli; Simon Levy (the Urban Community Forestry Project in London); Sarah Lloyd (NTFPs in the Boreal region); Willie McGhee (the Borders Forest Trust in Scotland); Yorgos Moussouris and Pedro Regato (NTFPs in the Mediterranean region); Mark Poffenberger; Maria João A. Pereira (community involvement in fire prevention in Portugal); Pauli Wallenius (public participation in Finland); Pier Carlo Zingari (Italian Forest Consortia).

I am extremely grateful for the artistic and careful work of Laura Rosenzweig, who constructed the maps; Tim Davis for the excellent layout; and Hild Glattbach from WWF International for the scanning of photographs and slides. Many thanks are also due to WWF International, DFID, and the Ford Foundation for funding this profile.

A very special thank you goes to Simon Rietbergen, who initiated the work of the European profile, and supported its progress with unfailing enthusiasm. His contributions at both conceptual and practical levels have been invaluable. Many thanks also go to Ursula Senn who helps administer the Working Group on Community Involvement in Forest Management (WG-CIFM). In the case of the European Profile, she provided assistance in establishing contracts and arranging workshops for the core team, with her usual good humour and remarkable efficiency.

Many examples of community involvement in forest management in Europe were visited and reviewed during the course of this work. I would like to thank many groups for their time and warm hospitality. Sadly, not every example can be included here. However, the pioneering work of so many individuals and groups who are working for the benefit of people and forests, in diverse ways, throughout Europe should be acknowledged. Their vision and experiences are invaluable and a great inspiration to us all.

ACRONYMS

BDG	BioRegional Development Group
BFT	Borders Forest Trust
BP	British Petroleum
DIY	"Do-It-Yourself"
CAP	Common Agricultural Policy
CEEC	Central and Eastern European Countries
CEPF	Confederation of European Forest Owners
CIFM	Community Involvement in Forest Management
CIS	Commonwealth of Independent States
DFID	Department for International Development (formerly ODA)
EC	European Commission
ECE (UN-ECE)	United Nations Economic Commission for Europe
EEA	European Environment Agency
EOMF	European Observatory of Mountain Forests
EU	European Union
EFTA	European Free Trade Association
FAO	Food and Agricultural Organisation of the United Nations
FF	Ford Foundation
FPS	Forest and Park Service (Finland)
FSC	Forest Stewardship Council
IFBWW	International Federation of Building and Wood Workers
IFF	Intergovernmental Forum on Forests
ILO	International Labour Organisation
IWGIA	International Work Group for Indigenous Affairs
IUCN	The World Conservation Union
NGO	Non-Government Organisation
NTFP	Non-Timber Forest Product
ODA	Overseas Development Administration (UK) (now DFID)
ONF	Office National des Forêts (France)

PEFC	Pan European Forest Certification
RSPB	Royal Society for the Protection of Birds
SBB	Staatsbosbeheer (Netherlands)
SEK	Swedish Kroner
SFM	Sustainable Forest Management
SSR	Swedish Saami Association
SWT	Scottish Wildlife Trust
TRN	Taiga Rescue Network
UK	United Kingdom of Great Britain and Northern Ireland
UNCED	United Nations Conference on Environment and Development
UN-ECE	United Nations Economic Commission for Europe
UNFCCC	United Nations Framework Convention on Climate Change
UNRISD	United Nations Research Institute on Social Development
UNEP	United Nations Environment Programme
WCMC	World Conservation Monitoring Centre
WG-CIFM	Working Group on Community Involvement in Forest Management
WGS	Woodland Grant Scheme (UK)
WWF	World Wide Fund For Nature (formerly World Wildlife Fund)

TABLE OF CONTENTS

LIST OF FIGURES

LIST OF TABLES

LIST OF BOXES

LIST OF PLATES

Figure 1

REGIONAL MAP OF WESTERN EUROPE

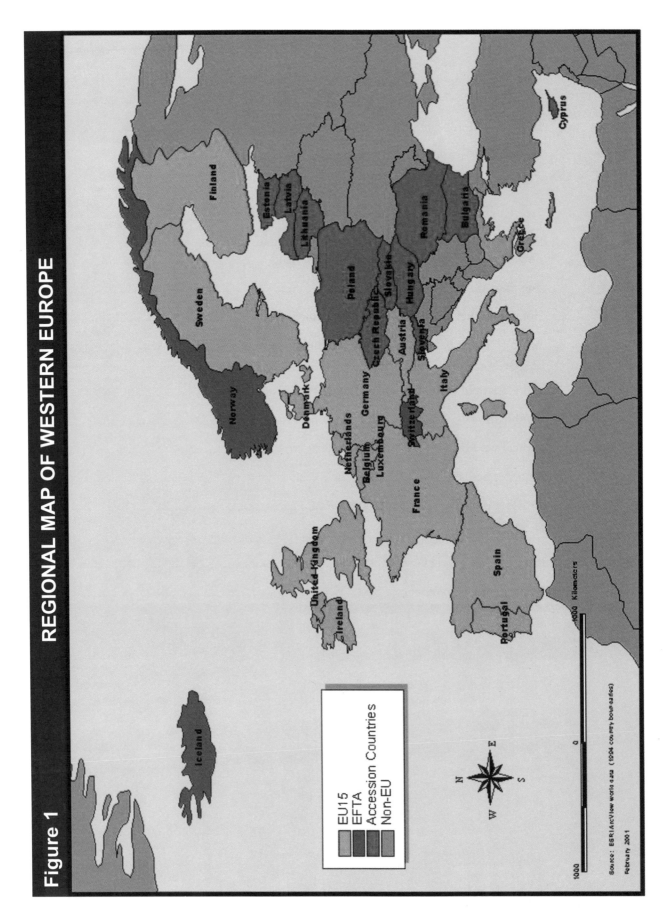

Part I

INTRODUCTION

OBJECTIVE OF THE EUROPEAN PROFILE

The objective of this profile is to explore, and learn from, the *diverse* and *changing* nature of community involvement in forest management (CIFM) in Western Europe. The profile also provides opportunities for examining the adaptations of CIFM to industrial contexts, and for envisioning the contributions of CIFM to sustainable futures.

EUROPEAN COUNTRIES COVERED IN THE PROFILE

The profile focuses on Western Europe, see Table 1, and draws mainly on experiences from countries within the European Union (known as the EU 15).[1] What constitutes Europe is a debatable matter and is not a subject of common agreement. The countries in Western Europe are characterised by immense diversity – of cultures, languages, histories, land ownership patterns, national policies and legislation, as well as climates, soils and forests. Nevertheless, based on broadly similar economic and political conditions today, European states are attempting to build a common regional identity. With the exception of Norway and Switzerland, all the countries below are within the political grouping of the European Union, and thus share a similar macro-policy environment.

Table 1	COUNTRIES AND POPULATIONS REPRESENTED IN THE PROFILE		
Country	**Population (million)[2]**	**Country**	**Population (million)[2]**
Austria	8	Luxembourg	0.5
Belgium	10	Netherlands	16
Denmark	5	Norway*	4
Finland	5	Portugal	10
France	59	Spain	40
Germany	82	Sweden	9
Greece	11	Switzerland*	7
Ireland	4	United Kingdom	59
Italy	57		
			* Not EU members

INTRODUCING THE FORESTS

Today there are about 136 million hectares of forests within the EU 15 countries, covering some 36% of the land area.[3] Trends indicate that there has been a 10% increase in forest cover over the last thirty years.[4] Forest and woodland cover varies from country to country, see Table 2 below. Apart from small pockets of natural forest, much of Europe's forest area is now managed and is subject to human activities. Forest composition has changed as a consequence, and more recently intensive management has led to a more homogenous and artificial forest. Some 68% of forestland is owned by 15 million private owners.[5] The vast majority of holdings are small with an average forest holding of about 10 hectares. Public ownership accounts for some 32% of the forestland in Europe. State forests are the principal forms of public ownership, although in several European countries, sizeable areas of forest are also owned by municipalities, communes or local governments.[6] Despite the diversity which characterises Europe, forest management has undergone a number of broad transformations common to the region as a whole. These have included expanding resource pressures; the growth of scientific industrial forestry;

and the more recent interest in multi-purpose, sustainable forest management. These transitions have a number of implications for understanding CIFM in Western Europe today. The development of the industrial forestry paradigm has typically been at the expense of rural communities and ecosystems, while more recent developments provide new opportunities for reconnecting people and forests. These transformations are discussed in more depth in Part II.

CHANGING REGION, CHANGING PEOPLE-FOREST RELATIONS

CIFM in Europe provides a lens on a *changing* region. Today, people-forest relations in Europe are influenced by changes affecting most of the western world – the globalisation of wood product markets; the effects of the intensification of forestry practices; urbanisation; changing policies and patterns of forest governance; and a growing environmental awareness. Although uneven in character, many social institutions are undergoing a profound transition from the unifying traditional values; beliefs and structures of 19th and 20th century modernity, with its industrial forestry standards, to a newly emerging era characterised by its emphasis on *sustainability*[8]

Table 2   FOREST COVER AND OTHER WOODED AREA AS A PERCENTAGE OF NATIONAL LAND AREA IN EUROPEAN STATES[7]

Country	Percentage of Forest and other Wooded Area Cover	Country	Percentage of Forest and other Wooded Area Cover
Ireland	9	Italy	37
Netherlands	10	Portugal	38
United Kingdom	10	Norway	39
Denmark	12	Austria	47
Belgium	22	Greece	50
France	31	Spain	52
Germany	31	Sweden	74
Switzerland	32	Finland	75
Luxembourg	34		

Figure 2

LAND COVER IN WESTERN EUROPE

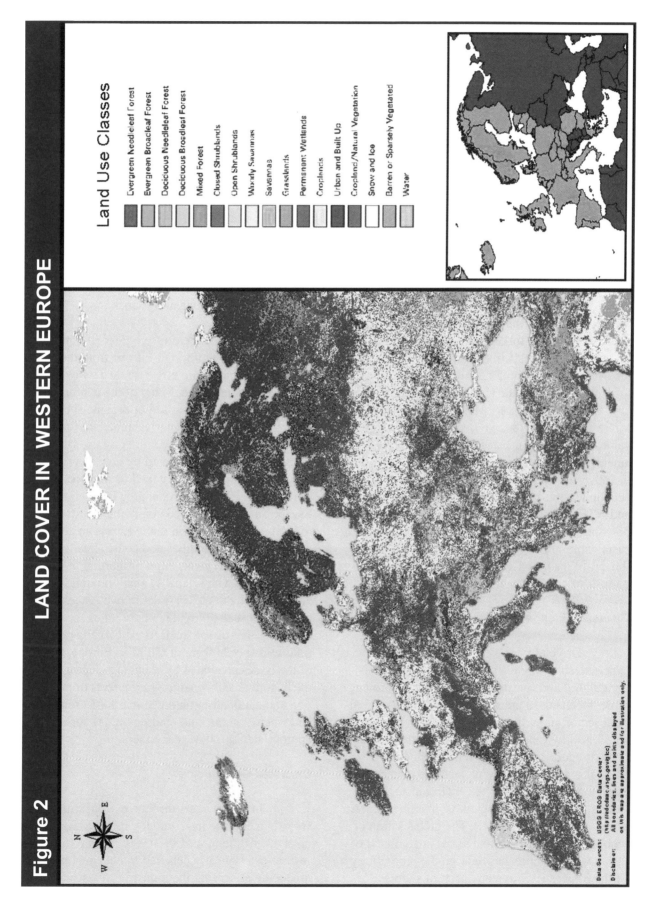

Land Use Classes

Evergreen Needleleaf Forest
Evergreen Broadleaf Forest
Deciduous Needleleaf Forest
Deciduous Broadleaf Forest
Mixed Forest
Closed Shrublands
Open Shrublands
Woody Savannas
Savannas
Grasslands
Permanent Wetlands
Croplands
Urban and Built Up
Cropland/Natural Vegetation
Snow and Ice
Barren or Sparsely Vegetated
Water

Data Sources: USGS EROS Data Center
(http://edcdaac.usgs.gov/glcc)
Disclaimer: All boundaries, lines and points displayed
on this map are approximate and for illustration only.

and plurality of values. As we shall see, CIFM in Europe is a mirror to these influences and changes, reflecting traditional, industrial-urban as well as new era concerns. This changing context generates a number of issues for CIFM in Europe, which are outlined briefly below, explored in greater depth within the profile, and returned to in the final chapter.

CHANGING ROLE OF FORESTS

Like elsewhere, rural peoples in Europe once depended on forests for an enormous range of products essential to their livelihoods. As sources of nutrients, forests, woodlands and hedgerows were also integral components of traditional agricultural systems. However, the social transformations of the last 300 years have resulted in a wide-scale de-linking of communities and forests, and of forests from agriculture. These days the average citizen no longer depends on forests for livelihoods, and the traditional roles of forests have all but disappeared. Many community forestry initiatives today reflect a new set of values of a largely urbanised population with amenity and recreational needs and objectives. On the other hand, the demands to revitalise local economies, and the possibilities of using timber and NTFPs in innovative ways, highlight the potential of a new era of people-forest relations for sustainable futures. One of the issues addressed by this profile is how CIFM in Europe is helping pioneer new roles for forests in society.

CHANGING ECONOMIC CONTEXT

Driven by market competition, Europe has experienced a long history of intensification of agricultural and forestry systems, resulting in large scale, mechanised practices. Recently, the agricultural sector, in particular, has been heavily subsidised by funds from the EU. While this has helped secure and stabilise food supplies, it has had a profound influence on land use patterns, often to the detriment of the environment, while failing to reverse the progressive marginalisation of many rural communities. The globalisation of wood product markets, and increasing volume of cheap wood imports has reduced the profitability of the forest sector in many parts of Europe.

However, the same industrial economic context has also provided opportunities for some communities to adapt forestry practices to urban society, providing many economic and social benefits to local groups. Contemporary developments, such as agricultural policy reform, new economic incentives, and product development are now helping to catalyse new initiatives throughout the region. Some of the questions generated by this changing context are how broader economic transformations are affecting people-forest relations in Europe; how some communities have adapted to industrial economic contexts; and what kind of incentives are needed to help catalyse and support CIFM in the region today.

CHANGING LAND TENURE AND RIGHTS

The liberal ideas of progress and modernisation which swept through Europe from the 17th century had a profound effect on land ownership and access to resources. Many of the commons were enclosed and privatised or became subject to different types of public and commercial forest management. Traditional community usufruct rights were usually liquidated by new tenure regimes, and are still threatened in some part of Europe today. However, changing land tenure in Europe has also produced commune forests (decentralised public forests) in many European countries, with special opportunities and constraints for local involvement, and a very high proportion of private forest owners with relatively small holdings. Their efforts to collaborate, in order to overcome diseconomies of scale, provide opportunities for exploring the role of small forest owner associations as a special form of CIFM in Europe. Some of the issues generated by changing land rights are the historical and contemporary effects of privatisation and nationalisation on people-forest relations; and whether some land tenure regimes appear more appropriate for CIFM than others.

CHANGING GOVERNANCE

In recent years discourses concerning forest governance have undergone considerable changes. At the turn of the 1900s, public forest institutions typically behaved like technocratic aristocracies serving the industrial forestry para-

4

digm.[9] However, in recent years there have been important moves towards multi-purpose forest management, along with growing emphasis on participatory decision making and, in some places, experiments with co-management systems.[10] Some of these changes are in response to the changing role of forests and the wider economy, but others reflect deeper changes in the character of western democracy itself, which is increasingly influenced by more interactive publics asserting their diverse needs and aspirations. Traditional community institutions have also been encouraged to adapt in the face of wider social changes; and private forest owners have had to address the question of other stakeholders' interests in forest resources. This profile is interested in whether changing policies and agencies in Europe are actually providing more opportunities for new patterns of CIFM; what kind of institutions provide effective and robust forms of CIFM; and how the rights of private forest owners can be balanced with interests of other actors.

In the last few years, Europe has witnessed quite significant shifts in environmental awareness – affecting the choices and behaviour of many sectors of society including public forest managers, wood processing industries and ordinary citizens. The legacy of the industrial, heavily mechanised forest management is just short of disastrous in many parts of Europe, threatening forest landscapes, ecosystems and species. Many new examples of CIFM in Europe have been shaped by concerns for the environment and sustainability, and may be considered sustainability pioneers, promoting an integration of social, environmental and economic objectives and values. Environmental concerns have also generated interest in the role of forest certification as a means to promote sustainable forest management. An important issue within Europe and elsewhere is whether CIFM makes any contributions to the conservation of biodiversity and landscapes.

Whether Europe will, or can, move into a new era of sustainability characterised by integrated social, economic and ecological values and objectives is difficult to predict. Social transitions

are always uneven, and shaped by contradictions and struggles. This profile argues that CIFM promises to play its part in this broader transformation, but is affected by the same struggles. Powerful institutional factors, and social power relations – both motors and inhibitors of change will shape its evolution. Some of these relations are introduced in this chapter and characterised in more detail later.

PEOPLE-FOREST RELATIONS: A DIVERSITY OF EXPERIENCES

The changing regional context, outlined briefly above, provides a key to understanding the diversity of people-forest relations in Europe today – as well as its possible contradictions. This profile presents a spectrum of people-forest relations, based on the experiences of different social groups, in a way that will hopefully capture something of the old, new and historically neglected forms of CIFM.[11]

RURAL GROUPS

In several parts of Europe, CIFM is based on ancient common forest property institutions.[12] These systems typically predate later municipal boundaries and structures, and continue to survive as viable tenure systems, providing community and environmental benefits. Several of these institutions have been strongly defended in the face of centralisation of state management of forests; privatisation; commercialisation of resources and urbanisation. Although not widespread, these cases are particularly interesting because they reveal important lessons on how communities have successfully evolved and adapted to political and economic changes in Europe. This profile explores several examples of older collective forest systems: the *Val di Fiemme* in northern Italy, the *Baldios* in Portugal. A further case study on the *Swedish Forest Commons*, indicates how they have adapted to urban, industrial society.

Some community forestry initiatives in rural areas are of more recent origin. The case study on the *Crofter Forestry* in Scotland illustrates the struggle of self-mobilised rural communities to obtain economic, social and ecological benefits from trees and forests for the first time. These initiatives can be appreciated better by knowledge of the long history of community disenfranchisement from forest resources – discussed in Part II. A further case study from the *Borders Region* in Scotland also looks at another community forestry initiative with important recreational and conservation objectives. The amenity functions of forests are increasingly important to rural and urban people in Western Europe. Self-mobilised forms of CIFM typically depend on policy changes (although they may also help prompt policy change); grants and subsidies; strong local leadership; coalitions of interested groups, and tough negotiations with government bodies to obtain forests for local and wider community benefits.

URBAN GROUPS

The demographic transition in Europe with its associated patterns of (de) industrialisation and urbanisation has laid the basis for new interest in the role of small woodlands in some urban areas. Since the late 1980s there has been a wave of local NGO initiatives seeking to promote environmentally and economically sustainable livelihoods and to rebuild communities, based on the production and trade of wood products. Sustainability pioneers, such as the *Bioregional Development Group* in London, UK, illustrate some of the opportunities and constraints of community involvement in urban forestry in a newly emerging post-industrial context.

INDIGENOUS PEOPLES

Indigenous peoples can encounter special problems of involvement in forest management in liberal democracies. The *Saami* – an indigenous peoples of Scandinavia, have traditionally been nomadic or semi-nomadic hunters who have relied on winter reindeer grazing within the boreal forests. In recent years their customary rights and land use practices have come under increasing pressure, particularly from commercial logging practices, and loss of access to some forests. The case study of the Saami in Sweden indicates how their customary rights of access and use of forest resources are being challenged by some small forest owners

through the courts, but supported through other participatory initiatives such as FSC certification. The outcome of these processes will have a direct impact on their way of life and forest resources.

FOREST OWNER/PRODUCER ASSOCIATIONS

Small forest owners – the largest group of forest owners in Europe – have developed alternative collective strategies to overcome the disadvantages of their small holdings. Many are members of forest owner/producer associations, which were often set up at a local level by dynamic community leaders. Small-scale producers worldwide often have a rough time in the market, as individually they have little influence on how prices for their products are set. Small-holder forestry is no exception to this, with the exception of some cases, such as Scandinavia. The profile contains a case study to explore how *Small Forest Owner Associations* have evolved; obtained a stronger position in negotiations with industry; how government forest policy and institutions have favoured such organisations. It also discusses forest certification issues facing small forest owner associations today.

FOREST WORKER UNIONS

Forest workers are often long-established members of rural communities, but the role of forest worker unions is frequently neglected in analyses of community involvement in forest management. Wood workers are frequently involved in day-to-day forest management, and their contribution to community involvement in forest management should not be underestimated. Unions of building and wood workers prevail through Europe, and are particularly strong in Nordic countries, Germany and the Netherlands. The unions represent their member"s interests, particularly for better wages and working conditions, and more recently for more sustainable logging practices. Unions also represent the interests of migrant forest workers.

ENVIRONMENTAL NGOS

Several environmental NGOs are supporting the role of communities in forest conservation, and have helped catalyse and promote local projects for conservation and socio-economic benefits. For example, a case study on *NTFPs in the Mediterranean* region illustrates how the WWF programme currently targets the conservation of significant Mediterranean forest areas through the promotion of rural community economic development by the sustainable management of NTFPs. This study highlights the important role NTFPs can play in sustaining rural groups as well as important landscapes.

COMMUNE OR MUNICIPALITIES

Commune or municipal forests are forests owned by the local governments, and are a feature of several European states, including Belgium, France, Luxembourg, Portugal, Spain, and Switzerland. In principle, decentralized administration of forests provides great potential for responding to and involving communities in forest management, and providing local benefits. The case study on *Commune Forests in France* illustrates the history and functioning of such commune forests indicating their particular strengths and weaknesses for CIFM.

PUBLIC FOREST AGENCIES

Like other regions, Europe is also undergoing profound policy changes which challenge the old professional technocratic styles of forest management and prompt more open and participatory processes of management. Forest management in Western Europe is increasingly influenced by more interactive publics and various interest groups, with distinct and often conflicting values, perceptions, objectives, knowledge and claims to forest resources. The profile presents European experiences from Finland, Denmark and Switzerland, illustrating how government forest services are adapting to more pluralistic forms of planning and management, and outlining some of the opportunities and constraints for new forms of CIFM.

CONCEPTS AND DEFINITIONS

Many concepts and terms used in this profile are outlined in the Glossary. However, the main concepts of *community, involvement, forest*

and *management* are defined briefly below.

COMMUNITY

The profile uses the term community[13] to denote self-defined, formal and informal, rural and urban forest user groups with shared values, knowledge and interests in forest management. This perspective includes *communities of interest*, which are not necessarily defined by location. These interests may include:

◆property, user and access rights

◆livelihoods based on the production of timber and non timber products

◆employment

◆cultural identity

◆leisure and recreation

◆biodiversity conservation and ecological restoration

In the past, the term community was often misleadingly used to refer to groups of people from the same *locality*. It is now widely recognised that groups of people may live in the same area but relate to forests in quite different ways. Such "communities" are not always homogenous groups, but can be differentiated along many axes such as gender, age, ethnicity, cultural values, knowledge, access to and control of resources like land, labour and capital and so on. These social differences shape access to resources, and have a profound influence on who benefits from and who bears the costs of forestry activities. In contrast, groups with shared values, interests and objectives may live dispersed over large geographical areas, and perhaps use resources only seasonally, such as the *Saami* in Northern Scandinavia. Problems with the notion of community have given rise to other concepts such as user groups, interest groups, stakeholders, coalitions and so on. See the Glossary for brief explanations.

The profile includes, but is not limited to the meaning of community forestry as used in the sustainable development and natural resource management literature, which often implies *rural development forestry* by, for and with local groups.[14] While this profile has a strong commitment to forestry which serves the needs of politically and economically marginalised rural peoples – and agrees that forests can often be most effectively protected by those who live near them and depend on them for their livelihoods – it acknowledges that a highly urbanised and industrialised region such as Western Europe provides a special context for CIFM. A broader view of community involvement in Europe can help highlight historically neglected forms of participation, and lend support to newly emerging democratic trends in public and private forest management.

INVOLVEMENT

This profile uses the terms "involvement" and "participation" synonymously. There are numerous ways to define and understand patterns of *involvement* in forest management, but it usually means the ability of individuals or groups to influence and share control over initiatives and the decisions and resources which affect them.[15] Patterns of involvement are subject to social and historical contexts and change over time. They may be shaped by one or more of the following:

◆*Tenurial Systems.* Some types of *common forestry property* ownership confer traditional decision making rights to their owners or representatives. These rights can be inherited or claimed in a variety of ways, and are discussed in more detail in the case studies from the *Val di Fiemme* in Part V.

◆*Membership of an Organisation.* Involvement in decisions and resources may be based on the rights conferred by organisational membership. This type of involvement is discussed in examples of *Small Forest Owner Associations* in Part V, and *Forest Consortia* – which are partnerships of private and public forest owners.

◆*Political Processes.* The meaning of involvement must also be attentive to political processes, that promote greater public participation in state forest management. Many new forest policy initiatives are deliberately encouraging new patterns of multi-stakeholder participation in forest planning

and management – moving forest management away from its traditional commercial objectives towards multiple objectives based on wider public consultation. These are discussed in Part IV.

One recent definition, which captures the contemporary meaning of participation as a political process, describes it as a "voluntary process whereby people, individually or through an organized grouping, can exchange information, express opinions and interests, and have the potential to influence decisions or the outcome of the matter at hand".[16] It can take place at all institutional and geographical levels – local, regional, national and international. Public participation in forestry should be seen as a *process*, which:

◆is inclusive of many stakeholders

◆based on voluntary participation (except where a legal requirement specifies otherwise)

◆should not conflict with legal provisions in force – particularly ownership and user rights

◆is fair and transparent to all participants, and follows agreed basic rules

◆is based on participants acting in good faith

◆does not have a predetermined outcome

Various scholars have produced typologies to help clarify and distinguish various *degrees* of involvement. For example, Arnstein (1969)[17] outlines eight levels of public participation: 1-manipulation, 2-therapy, 3-informing, 4-consultation, 5-placation, 6-partnership, 7-delegated power, and 8-citizen control. Typologies can help highlight the complex and ambiguous ways in which the terms can be interpreted. Indeed, the terms may be used as policy narratives to legitimise the activities of particular actors, and can disguise what is actually taking place. Careful consideration must be given to discourses of participation, and how language is being used. As suggested, processes of participation and involvement are influenced by a number of formal institutions, such as tenure regimes, governance structures and processes. But, they are also shaped by hidden and deeply embedded social norms or informal rules, which have a profound effect on social relationships.

FOREST

Forest is land with a tree crown cover of more than 10% and an area of more than 0.5 ha.[18] The trees should be able to reach a minimum height of 5 metres at maturity in situ. Forest may consist either of closed forest formations where trees of various storeys and undergrowth cover a high proportion of the ground; or of open forest formations with a continuous vegetation cover in which trees crown cover exceeds 10%. Other wooded land consists of land with a tree crown cover of 5-10% of trees able to reach a height of 5 metres at maturity in situ; or a crown of more than 10% of trees not able to reach a height of 5 metres in situ, such as dwarf or stunted trees and shrub and bush cover.

MANAGEMENT

In this profile the term management is used in both narrow and broad senses. It includes, but is not limited to, the operational plans which guide activities within the forest. The term "management" often refers to a forest or wood which is managed in accordance with a formal or informal plan applied regularly over a sufficiently long period (five years or more). The management operations include tasks to be accomplished in individual stands or compartments during the given period.[19] However, the term management can also refer to the patterns of governance[20] which influence the style or culture of decision making. For example, several of the examples cited within the report identify new ways in which management planning and decision making are undertaken by public forest administration, allowing stakeholders to influence decisions and actions. Management may also be based on experience or traditional knowledge, which is characteristic of some of the older forms of community forestry in Europe.

The profile also refers to the concept of *sustainable forest management* in Europe, which has motivated many contemporary community initiatives. There have been a number of definitions of sustainable forest management agreed by international political fora in the last 10 years. Most definitions refer to at least three major dimensions of sustainability: ecological, social and economic. One of the earliest definitions is from "The Forest Principles" produced at UNCED in 1992[21]: "Forest resources and forest lands should be sustainably managed to meet social, economic, ecological, cultural and spiritual human needs of present and future generations. These needs are for forest products and services, such as wood and wood products, water, food, fodder, medicine, fuel, shelter, employment, recreation, habitats for wildlife, landscape diversity, carbon sinks and reservoirs, and other forest products". Sustainable forest management is linked to the broader concept of *sustainability*, which is defined in the glossary.

COMMUNITIES OF INTEREST AND POWER IN WESTERN EUROPE

The discourse around "community" has evolved considerably in the sustainable natural resources literature in the last 30 years, and has become more politicised. It is clearly important to identify the different interests in forests, as well as the sources of power and scales of influence of various groups, along with the different means people use to achieve their interests. Table 3 below illustrates a typical, if generalised, range of forest user groups in Western Europe. It is acknowledged that these are very broad groups, which may contain important intra-group differences. Boundaries between groups can also be very fluid and may change over time. The purpose of the Table is to help tease out some general differences in Europe and give a preliminary view of power relations which are implicit in several of the case studies.

In general, the table reveals the importance of many non-market goods and services provided by the forest in Western Europe to several groups, such as recreation, cultural heritage and conservation values. It also indicates the overlap between the demands of different groups. These interests are sometimes incompatible. For example, conflicts can arise between indigenous peoples and small forest owners;

Table 3	FOREST INTEREST GROUPS AND POWER IN WESTERN EUROPE			
Group	**Interests**[22]	**Source of Power**	**Scale of Influence**	**Means to Achieve Interests**
Rural/Urban groups and individuals	Access to and use of forests Employment Income Consumption of wood and non-wood products Recreation and leisure facilities Symbolic value of trees and forests	Variable, but generally local Voting power	Local	Local action Networking Coalitions created/joined to extend influence beyond local
Indigenous people	Access to and use of forests Recognition of traditional usufruct rights Old-growth forests Cultural recognition	International recognition, but generally politically and economically disadvantaged	Local Regional International	Local action Political lobbying Creating/joining coalitions to extend influence beyond local
Associations of Small Forest Owners	Private property rights Income/Profit Conservation of cultural, environmental and economic heritage Privacy	Forest ownership Economic importance Government support	Local Regional National Pan-European	Local action Networking Political lobbying
National Public Forest Agencies	State forest property rights Income/Profit Conservation of cultural, environmental and economic heritage Provision of public infrastructure and services	Forest ownership Political Economic Administrative	Local Regional National	Implements national policy

Table 3 continued...	FOREST INTEREST GROUPS AND POWER IN WESTERN EUROPE			
Group	Interests[22]	Source of Power	Scale of Influence	Means to Achieve Interests
Commune and Municipalities	State forest property rights Income/Profit Conservation of cultural, environmental and economic heritage Provision of public infrastructure and services	Forest ownership Political Economic Administrative	Local Regional National	Implements national and sub-national forest policies
Forest Workers and Unions	Sustainable employment Income Health/safety Job satisfaction	Variable Economic Political	Local Regional National International	Collective bargaining Lobbying for workers' rights at local, national and international levels
Conservation and Environmental Groups	Protection of rare, endangered species and habitats Ecologically sound forest management and restoration Environmental public education	Financial support from individuals, governments and industry Scientific knowledge Public support Wide range of partnerships	Local Regional National International	Financial and technical support to local groups Lobbying for policy change at national and international levels Public education, communications
Sport and Recreation Associations	Access to and use of forest Sport, recreational, educational facilities and infrastructure	Public support Financial support from governments and (local) businesses Partnerships	Local Regional	Lobbying
Private Forest Companies	Private property rights Profit Satisfy shareholders	Economic importance Political support	Local Regional National International	Political lobbying

Table 3 continued...	FOREST INTEREST GROUPS AND POWER IN WESTERN EUROPE			
Group	**Interests[22]**	**Source of Power**	**Scale of Influence**	**Means to Achieve Interests**
Hunting Associations	Access to and use of forest High game density for meat and trophies Wildlife habitats	Public support Influential individuals	Local Regional National	Lobbying

between hunting associations and conservation groups, public forest agencies and rural groups, workers and industry and so on. The profile thus acknowledges conflicts of values between different *communities of interest* in forest resources in Europe.

The table also outlines the relative degree of power of different groups, indicating that some groups are more successful than others in achieving their interests or getting their voices heard. The unequal power relations are rooted in different sources of power, such as land and forest ownership, economic importance, political function, etc; which in turn shape the scale of influence and the methods groups use to achieve particular interests. An analysis of power relations can also help clarify group relationships and intersecting layers of interest, such as government support to small forest owners, or environmental NGO support to rural groups, etc. The relative strength of these social networks also provides opportunities or constraints to particular types of CIFM. On a practical level, an analysis of power relations can also help inform strategic alliances for change. The experiences of several of these groups, and various coalitions of interests are explored in more detail in Part V.

ORGANISATION OF THE PROFILE

◆The history of people-forest relations in Europe, and the implications of current social and forestry trends for CIFM is explored in Part II.

◆The natural forest ecosystems in Western Europe, and some of the problems facing European forests today are described in Part III.

◆European-level data on institutions shaping CIFM, and different national contexts are outlined in Part IV.

◆Case studies illustrating a diversity of people-forest relations in Western Europe are presented in Part V.

◆Part VI discusses opportunities and challenges for CIFM in Europe in the 21st Century and draws out the main lessons for different groups.

◆Part VII makes recommendations and presents a vision of CIFM for sustainable futures in Europe.

Notes

[1] There are several accession countries to the European Union, as shown in Figure 1.

[2] UN-ECE/FAO (2000): *Forest Resources of Europe, CIS, North America, Australia, Japan and New Zealand.* Geneva Timber and Forest Study Papers, No.17. Main Report: pp.62-63. United Nations: New York and Geneva.

[3] UN-ECE/FAO (2000): op.cit.

[4] Stanners, D. & Bourdeau, P. (Eds) (1995): *Europe's Environment. The Dobris Assessment.* European Environment Agency: Copenhagen.

[5] UN-ECE/FAO 2000: op.cit.

[6] From the available statistics it is difficult to estimate the proportion of common or communal forest properties, see glossary for definition. These data are either incorporated into private or public ownership statistics.

[7] UN-ECE/FAO 2000: op.cit.

[8] See glossary for definition.

[9] Kennedy,J.J; Dombeck, M.P. & Kock, N.E. (1998): "Values, beliefs and management of public forests in the Western World at the close of the twentieth century". *UNASYLVA* (49):192: 16-26.

[10] See glossary for definition.

[11] Social diversity can be organised in several different ways, according to group, management objectives, rural/urban locations, historical contexts, and so on. In the European context it should be borne in mind that many of these categories might be artificial. Groups typically have multiple management objectives that extend beyond rural development or conservation. On close examination it will be found that a range of partners may be actively involved in so called community initiatives, including local people, representatives from environmental NGOs and public forest agencies. Urban groups frequently use rural forests; small forest owners increasingly live in urban areas and so on.

[12] Common property institutions or regimes (CPRs) are defined in the glossary. CPRs are found in Northern Italy, Portugal and parts of France and Switzerland.

[13] There are many excellent discussions on 'community'. See:
- Agrawal, A. (1997): *Community in Conservation: Beyond Enchantment and Disenchantment.* CDF Discussion Paper. Conservation and Development Forum, University of Florida, USA;
- Guijt, I. & Shah, M.K. (Eds) (1998): *The Myth of Community. Gender Issues in Participatory Development.* Hants: Intermediate Technology Publications;
- IDS Bulletin (1997): *Community-Based Sustainable Development: Consensus or Conflict?* University of Sussex: Institute of Development Studies No.28 (4);
- Li, T.M. (1996): "Images of Community: Discourse and Strategy in Property Relations". *Development and Change* 27: 501-527;
- Peters, P. (1996): "Who's Local Here? The Politics of Participation in Development". *Cultural Survival* 20 (3): 22-25.

[14] According to FAO's 1998 community forestry strategy: "the vision of community forestry is a partnership of communities, forest agencies and other stakeholders working together for sustainable forest management and rural development". It supports the control, management and use of forests and tree resources by local communities. It explores the social, economic and cultural relationships between people and forests. It implies a decentralised and participatory approach to forest management, which assumes that the best stewards of the world's forests are the populations living in and around them". Rome: FAO Community Forestry Internet Site, 1998.

[15] Adapted from a World Bank definition of participation. Washington D.C: World Bank Rural Development Internet Site, 12.02.01.

[16] FAO/ECE/ILO (2000): *Public Participation in Forestry in Europe and North America.* Working Paper 163. Geneva: International Labour Office.

[17] Arnstein, S. (1969): "A Ladder of Citizen Participation".*Journal of the American Institute of Planners.* Vol 35:216-224.

[18] The Profile adheres to the general definitions of forest and other wooded land used in the UN-ECE/FAO Temperate and Boreal Forest Resources Assessment 2000, op.cit. unless otherwise specified.

[19] UN-ECE/FAO (2000): op.cit.

[20] See glossary for defnition.

[21] Robinson, N. (Ed) (1993): "Non-Legally Binding Authoritative Statement of Principles for a Global Consensus on the Management, Conservation and Sustainable Development of all Types of Forests. In *Agenda 21: Earth's Action Plan*. New York, London, Rome: Oceana Publications.

[22] Adapted from FAO/ECE/ILO (1997): *People, Forests and Sustainability: Social Elements of Sustainable Forest Management in Europe*. Geneva: International Labour Office: pp.10-13.

Part II

A BRIEF HISTORY OF HUMAN-FOREST RELATIONS

INTRODUCTION

Forests are cultural as well as ecological spaces. Each historical age reveals something about its values, social relations, human institutions and laws through the diverse ways in which forests are managed.[1] Forests embody our cultural values, just as trees, woodlands and forests play an important role in our cultural imagination.[2] This chapter briefly explores the changing nature of forest management and use in relation to significant social transformations within Europe, and their implications for CIFM. There is no single history of forest resources – or of CIFM – in Europe, but rather many different histories, conditioned by a wide variety of specific bio-physical, political and economic contexts.[3] However, for the sake of clarity, features common to many European countries are discussed below in historical sequence. Readers should bear in mind that in outlining general trends the course of events is never as simple and linear as it may appear.

NEOLITHIC AND BRONZE AGES: EARLY AGRICULTURAL CONVERSION

Europe became suitable for tree growth in the millennia after the last Ice Age (about 11,000 BC), and became covered with natural forests, sometimes called "wildwood".[4] It is estimated that some 80%-90% of Europe's land once had this original forest cover.[5] Most of the great wildwoods passed away in prehistory, although some remnants survived into historic times in

Europe, and are known of through legends. Deforestation was partly due to the changing climate, but principally to human activities such as land clearing for farming.

The original woodlands were first colonised by hunters and gatherers, but Palaeolithic and Mesolithic peoples made little impact upon the forests. Neolithic settlers arrived in the UK around 4000 BC and, through new practices (livestock rearing and settled agriculture), had significant impact on the landscape through converting large tracts of countryside to farmland or heath. Archaeological evidence, such as the Sweet Track, the wooden walkway built to cross the soft peat in the Somerset levels dating from 3900 BC, suggests that timber from different hardwood species were harvested from mixed coppice woodland, rather than from wildwood. Evidence suggests that Neolithic wildwoods were converted into managed woodland and hedges to provide a variety of useful products such as wattle for hurdles used in fencing and house and road building; fuel for cooking, heating, metal working and pottery.[6] Woodlands were also used as pasture.

Changing farming practices in the Bronze Age (1500 BC) produced a landscape characterised by more or less permanent pastures and arable fields with only short fallow periods. Wood-fuel use increased as pottery, metal working and later glass making became important. In Germany, repeated clearing by fire in the Bronze age led to the development of the first heath landscapes in northern regions. It is estimated that half of England had ceased to be wildwood by the early Iron Age (500 BC).[7]

CELTIC AND ROMAN PERIODS: ANCIENT RIGHTS AND RESPONSIBILITIES

Institutions regulating tree and woodland management were known during Celtic and Roman periods. In Ireland the Brehon Laws date back to the 7th century AD, and include tree listings and classifications, and details of penalties imposed for poor tree management. The Celts treated the woodlands as commons, held jointly by members of the tribe. There was no concept of private ownership or land transfer. Rights were balanced with responsibilities and obligations.[8] Greek and Roman authors from 700 BC to 300 AD recorded on-farm tree planting and management for subsistence purposes, such as coppicing techniques for firewood, fodder and poles and the role of multipurpose tree boundaries.[9] Social institutions underpinning woodland management also prevailed during periods of Roman control. Within Italy, forests and pasture lands were considered public property and were managed collectively, while agricultural land was privately owned. Communal use of public lands was supported by *usi civici* (traditional usufruct rights). They continue to exist to this day in many areas.[10] With the expansion of the Roman Empire about two thousand years ago, large amounts of timber were used for ship building to support the fleet and fuel for industries. Evidence from the UK indicates that the Romans imported forestry skills and relied on managed woodlands for their military iron works, buildings, bridges and ships, and for their baths, brick making, iron lead and glass works. With the fall of the Roman Empire, forests regenerated until renewed economic growth began to take its toll in the forests again after the 10th century.[11]

THE MEDIEVAL PERIOD: THE ROYAL ROOTS OF FORESTS

In the Middle Ages, forests were traditionally royal hunting areas – land on which deer were protected by special laws. Forests were not necessarily wooded places, but included villages, cultivated fields, pastures, areas of heath and woodland: "to the medieval a Forest was a place of deer, not a place of trees".[12] The word 'forest' only came to imply woodland several centuries later. In the Medieval period, royalty owned the deer but not necessarily the land or the trees. Most forests were also commons and had common rights dating from before they had been declared forests.

The concept of forests, and related laws, were introduced to Britain from the Continent by William the Conqueror. Royal forests were controlled by a hierarchy of forest officials, and subject to special kinds of laws administered by Forest Courts, which were notoriously severe, and used as a method for collecting revenue. The poaching of venison was the most serious crime, and resulted in large fines or imprisonment. Outside royal forests the landowning classes applied the principles of the royal forest to their own land. After the Norman Conquest in Britain in 1066, the saxon notion of land 'ownership' by working farmers was eliminated. It was replaced with the idea of land ownership through conquest, introducing the concepts of land possession and dispossession. Some claim that the emphasis on hunting, and the privileged status of private landowners at the expense of the community, dating from the medieval period, have helped shape attitudes and behaviours and have bequeathed several principles of land management which persist in Britain to this day.[13]

In northern Norway, rules governing the rights and duties towards forested Commons have an ancient history, dating from the ages of the great migrations, through the Viking age, and were transferred orally from one generation to the next. The early provincial assemblies, which pre-dated Christianity, were initially self-governing and independent of the Church and King, and could adapt laws as deemed appropriate in order to accommodate new demographic, technological or cultural developments. The ancient customary rules governing the commons were copied from the regional assemblies and became an important part of the first Common Law of 1274, at the time of the unification of the country. The rules of the commons in Norwegian Law survived in its original wording until January 1993, representing an unbroken tradition of oral and codified customary

In Britain, the feudal tenurial systems and the Royal forests permitted community rights to the woodlands and its products – rights which dated from time immemorial.[15] The commons were far from being patches of waste but provided valuable resources to the commoners, from landless squatters to the richest farmers. A typical wood common was grassland thickly scattered with trees and bushes. Wooded commons belonged to a landowner, usually lord of the manor, but the rights to use them belonged to commoners, while the soil, including mineral rights, belonged to the lord.

Grazing rights were the most important of the numerous common rights. They enabled people with little or no land to graze livestock, such as pigs, geese and horses. The number of livestock allowed by each commoner was limited, and extra animals were grazed on payment of a small 'fine'. Pigs were fed on acorns or beechmast in the autumn before being killed in the winter. The lord received one pig in ten for this common-right. Manorial courts did not always manage to preserve trees on commons due to severe grazing pressure. Wood was another important resource. Often the timber belonged to the lord, while the coppice was used by the commoners. A common practice was to pollard trees – to produce repeated crops of poles and fuelwood – to prevent the animals eating the regrowth. Commoners also had access to private wood and hedgerows for timber. Hedges and hedgerow trees have been sources of useful products in Britain for a thousand years.

The management of the commons was subject to numerous bylaws related to keeping the rights open, and to protecting animals. Local manor courts, often composed mainly of the commoners themselves, prepared, enforced and revised rules which regulated the commons. These did not always favour the lords' interests. Common law did not allow common land to be fenced off for private use without the consent of free tenants. The commoners rights facilitated a co-operative farming system which dated from earlier times and survived under Norman feudalism. By the 1300s many of the commons had disappeared in the United Kingdom. Either the grazing had turned them into heaths or grasslands, or they had been privatised. In the UK the enclosure of the commons, and private appropriation of land finally got underway with the Acts of Enclosure in the 18th century, and the Highland Clearances in Scotland, resulting in wide scale loss of common rights. Nevertheless, ancient common rights still persist in some areas, such as the New Forest within south-west England.

law over more than 1000 years. However, as discussed later, a number of piecemeal changes over the centuries has gradually eroded customary law regulations.[16]

"Commons Forest can everyone use, who lives there, to whom it belong from Ancient Times, as far as it is for the need of everyone for Firewood, for necessary Building Timber and for Farm Use. Who cuts more forest in Commons and burns fires without Permission for Clearing, as it is said, is to be charged the same as the one who cuts in another man's Forest". Law of King Christian V (1687) Book III, Chapter 12.

Despite ancient codified rights and responsibilities throughout Europe, forests continued to be widely depleted. France in the 12th century was the time of the *"moines-défricheurs"*, a term pertaining to monks who cleared the forests for cultivation. The word défricher, or clearing, aquired a positive connotation that it retains today in its figurative uses.[17]

EARLY INDUSTRIALISATION 16TH –18TH CENTURIES: FURTHER DECLINE IN FOREST RESOURCES AND COMMUNITY INVOLVEMENT

Many factors contributed to the decline of

forest resources throughout Europe from the 16th century: population expansion, the continuing conversion of forests to agricultural land, and use of vast quantities of charcoal for early industrial activities such as iron smelting; salt and glass works, and mining. In Ireland, industries such as iron smelting and glass works used substantial quantities of wood. One ton of iron required 2.5 tons of charcoal, the equivalent of roughly an acre of 5 year old oak coppice. By the late 18th century almost all the country's wood needs were imported.[18] In densely populated parts of Southern Sweden the forest became so depleted between 1650 and 1850 that cow dung had to be used instead of fuelwood in several places.[19] Early forest regulations often proved difficult to enforce.

In Belgium in the sixteenth century, the metallurgical industry was demanding so much charcoal to supply the forges that charcoal burners were blamed for deforestation. A Forest Edict on Woodlands and Forests was introduced in 1617 which forbade the creation of new forges, but the regulations failed to be implemented.[20]

Demand for high quality ship building timber also began to increase, as several European countries became merchant, naval and later imperial powers. In Italy, forests came under renewed pressure in the Middle Ages as the City states, such as Pisa, Genoa, Venice, began to develop and trade in the 14th century. At the high point of their development it is estimated that the combined

---

Box 2   **TIMBER BARONS AND TIMBER EXPORTS IN 18TH CENTURY GERMANY**[22]

Timber exports from the Black Forest in Germany increased in the mid 1800s, when cities along the Rhine began to rebuild and expand after long years of warfare. Holland, which rivalled England as a maritime and colonial power, made an enormous demand on timber. At the height of the golden age of shipbuilding in the Netherlands, in the 1600s, some 500 ships were built annually in Amsterdam and surrounding lands, requiring about 300,000 cubic metres of timber.[23] Demand for timber from Germany rose so steeply that the price of timber doubled in the margravate of Baden between 1750 and 1790. Timber from the Black Forest came to be known as 'Black Forest gold'.

Dutch trading associations began sending their representatives directly to Black Forest areas to purchase what came to be known as 'Dutch firs', giant logs up to 33 metres long and 48 cm in diameter at the smaller end. These were used in the construction of buildings and harbours, and also to make the piles which still support the foundations of older homes in Rotterdam and Amsterdam. Timber was rafted down the Rhine, sometimes in vehicles of astounding dimensions, of up to 300 metres long and 50 metres wide. As many as 500 oarsmen were often required to manoevre such rafts. These were often used to transport oak, which was worth ten times as much as the conifers but was too heavy to float. Timber from the Black Forest reached Paris via a network of canals in France. Rafting of timber came to an end at the end of the 1800s, with the advent of steam powered vehicles and the building of the rail network.

In some areas of the northern part of the Black Forest in Germany, local business men, such as inn keepers, keepers of post houses and civil servants, financed timber harvesting on public forestland on a large scale. They formed trading associations composed of merchants, backed by significant amounts of capital. These formed a core of 2 to 3 dozen 'timber barons'. Enormous profits were to be made very quickly in the timber trade with Holland. For example, between 1755 and 1809, the timber company in Calw paid out almost 2 million Gulden, with an annual return on investment of between 17 and 56 per cent. However, timber cutting on public forestland was often done hastily and with no attention to sustainable practices. In the Black Forest of Württemberg, one third of the public forest land was clear cut and left denuded by 1817. Privately owned farm-forests were more sustainably managed in southern parts of the Black Forest, in spite of the demand for Dutch firs.

PLATE 1: Women carrying fuelwood in the Canton of Tessin, Switzerland, in the early 1900s. (Courtesy of Küchli 1997.)

needs of their ship yards required some 18,000 cubic metres of wood per year. This resulted in the widespread adoption of forest inventories and management controls. [21]

RURAL FOREST USERS

Forests continued to play a central role in rural economies, providing fuelwood, animal fodder, pasture, and in some areas in providing food such as chestnuts. Many rural areas depended upon wood processing as the basis of their local economy. In 1850 the village of Bernau, near Freiburg in the Black Forest region of Germany, recorded that there were: 120 coopers (tub and barrel makers); 30 box makers; 12 spoon makers; 5 wood turners. The forest was often the primary source of livelihood for landless people. For instance, in the Bernese Oberland in Switzerland, up to a quarter of the families possessed no land of their own.[24] Landless people, often recent migrants, had only unofficial rights to forest resources. They obtained fuelwood by gathering the branches left behind by those who had common rights to use entire trees, and scraped up

leaves and soil from the forest floor to fertilise the potatoes they grew on tiny patches of leased land. They also kept goats, "the poor man's cow", which browsed in the forest.

URBAN PRESSURE ON FORESTS[25]

Cities like Berne in Switzerland acted as a "black hole", devouring enormous amounts of energy and raw materials from its surroundings. As early as the 14th century citizens sought to conserve their forests by means of strict regulations. Later, it was stipulated exactly who could use wood for cooking and heating, and in what amounts. Yet despite warnings against high consumption and waste, by about 1800 the citizens of Berne, who numbered some 12,000 at the time, were consuming around 50,000 cubic metres of wood per year. In order to satisfy its demand for energy, the City of Berne imported wood from the Bernese Oberland, and timber was rafted to the city down the River Aare. The District Forester at the time, repeatedly condemned official imports of timber by the City as a strategy to avert the so-called shortage of

19

wood in order to keep prices low. In his opinion, this artificial reduction in the price of a resource hampered energy-saving innovations, such as the use of improved stoves and ovens.

CHANGING PROPERTY RELATIONS

The expansion of modern states in the 17th and 18th centuries throughout Europe had a dramatic effect on property relations. The commons were frequently perceived as obstacles to a free market economy and economic growth, and the pervasive practice was to replace them with 'rational' forms of property rights: private and state property. New legislation was introduced to include village communities within newly established local authorities, such as the communes or municipal councils. This undermined traditional forms of community forest management. For example, in Spain, many lands were traditionally owned by parish communities or belonged to a given locality. During the 18th and 19th centuries, parish and/or municipal councils

usurped commoners' rights, claiming that the land belonged to the 'village council' rather than to the 'villagers'. Between the 1940s and 1980s many of these areas became sites of massive reforestation projects, under the control of the State forest administration. Fast growing conifer plantations have had negative impacts on employment, particularly from the loss of grazing, the landscape and the environment, leading to increased poaching on previously held commons, forest fires, soil erosion, habitat and biodiversity loss.[26]

A dominant discourse throughout Europe from the 18th century was to blame forest degradation on expanding rural populations and local farming practices such as fuelwood and fodder gathering. However, other interpretations suggest that the enclosure of the commons throughout Europe has not necessarily resulted in a net gain in productivity, but rather a redistribution of income in favour of the state or private landlords.[27] Others argue that problems of fuelwood and subsistence foraging in Europe should be interpreted within the context of the privatisation of the commons, resulting from an expansion of state control in rural areas, rather than destructive behaviour of rural people per se.[28] Perceptions of local people as the cause of problems, and of the commons as obstacles to 'progress' have supported state appropriation of land usually at the expense of rural livelihoods.

MODERNISATION 19TH – 20TH CENTURIES: THE EXPANSION OF FORESTS AND THE NEW TECHNOCRACY

PRESSURE ON FORESTS EASES OFF

During the eighteenth century dependence on fuelwood began to decrease as coal began to replace charcoal for commercial production and domestic use.[33] Although timber was rapidly devoured in the initial phase of railway construction, the great quantities of coal brought in by rail were a beneficial infusion that began to reverse the degradation affecting most forests in Europe. For example, in Switzerland coal had already become less expensive on the Bernese energy market than

wood by 1860, which it rapidly began to replace. Steam-powered ocean-going vessels and the expanding railway network provided the transport infrastructure that made it possible to obtain raw materials, cereals and fertiliser from overseas. The availability of new jobs in cities that developed around railway nodes caused many people – particularly those who belonged to the underprivileged classes – to migrate from rural to urban areas. One consequence of the industrial age was reduced pressure on the forest, allowing foresters to finally realise their visions of forest management, involving the introduction of new forest legislation and systems of management, modern forest administrations, forestry training schools and universities. Together, they have had a dramatic impact on the expansion of forest throughout Europe.

SUSTAINED YIELD FORESTRY

The economic value of timber helped establish the birth of forestry as a modern scientific discipline. The idea of a regulated and *sustained yield forestry* first prevailed in the 16th century in central Europe, but was further developed on the basis of research and experience in the

---

Box 4   **PRIVATISATION OF THE FORESTS FOR PROFIT IN SWITZERLAND**[30]

Liberal reforms extended to forestry in Switzerland in the 1800s. Forests were divided up in such a way that each owner could use or improve his own parcel of land in accordance with his own wishes, "unhindered by lazy, envious or ignorant fellow owners".[31] Despite the guarantee of equality to every citizen to forest resources in the new constitution in cantons like Berne, not every family obtained rights of use in the forest. In practice, members of the rural upper class succeeded in exercising their privileges, and the traditional customary rights which had long been tolerated were then abolished. In 1835 an organisation of landless citizens reported that both established and newly privileged social groups were claiming extravagant amounts of timber for personal use, and enriching themselves by selling it, and thereby causing long term ruin of the forests. In the Canton of Valais in the central Alps, clear cutting took place on a massive scale in response to external demand. Powerful local families who had controlled important Alpine trade routes for centuries were among the first to become involved in the Asian spice trade, and took advantage of their position to profit from the timber export trade. The authorities were too timid to bring charges against unrestrained felling and the state became the laughing stock of those who divided the forest up among themselves. At the same time, the denial of social justice for all made thieves of those who had been excluded from the new economic order. Pillaging of the forest now took place on a massive scale, increasing in direct proportion to the rise in the price of timber.

18th and 19th centuries.[34] In contrast to selective felling of trees, or the *Planterwald* system, the sustained yield system aims for a more controlled use of timber biomass. It involves dividing the traditional mixed forest into units and managing units to produce stands in which trees of the same age and species can grow and be felled together. Another target of the sustained yield concept is to afforest bare lands and to convert scrub vegetation into high forest. The resulting chessboard forestry techniques became legally binding in the German states during the period 1833-1900, and selective felling of trees was forbidden. The latter was described as an inefficient and uneconomical practice that violated all the rules of forestry. Many farmers in the southern part of the Black Forest in Germany reacted violently against such authoritarian attempts to control their activities, and numerous communes repeatedly petitioned the Government, which finally modified the ban on the *Planterwald* system.[35] Since the end of the

Second World War, the sustained yield concept has gradually been challenged by a broader idea of 'sustainability', with the object of maintaining diverse forest functions and services.

THE NEW FORESTRY TECHNOCRACY

The first academic forestry institutions were established in Germany at Göttingen and Tharandt at Dresden. They produced some internationally famous foresters with a world-wide reputation and who were employed in German, British and Dutch colonies. The first forest school in the UK was established at Durham in 1854, by the German forester Dr. Schlich. The Ecole de Nancy in France, founded in the 1850s, also offered international forestry training. The foundation of the British Colonial Service in India was helped with the training of 82 foresters at Nancy between 1867-75.[36] The eductation of Dutch state foresters started in 1883 at Wageningen.[37]

Box 5	TREES: A SYMBOL OF LAND DISPOSESSESSION IN IRELAND[32]

In Ireland, the English Tudor regime (1485-1603) adopted a policy of conquest, which involved confiscating Irish lands and transferring them to English settlers. In the late 16th century, Queen Elizabeth I of England ordered the destruction of Irish forests as a means of gaining greater control over the land. Later in 1793, the Dublin Society was formed and promoted a premium scheme for various afforestation projects of oak, ash, elm and pine, which lasted 40 years. It is estimated that between 1766 and 1806 some 25 million trees were planted. However, the main beneficiaries were the large landowners who were able to invest in the development of their lands. To many Irish tenant farmers trees were a symbol of land dispossession.

With the Act of Union in 1800, London replaced Dublin as the capital producing the phenomena of the 'absentee landlord'. Land rents were high to support the landlord's lifestyles, and tenant farming was characterised by tenurial insecurities, persecutions, evictions and food shortages. These hardships fostered the growing independence movement which had its more formal origins in 1793, and had a strong focus on land and land rights. The Great Famine of 1845-9 resulted in the death of one million people from hunger and disease, and prompted mass emigration of some 2 million people mainly to Canada and the United States. It stimulated greater political solidarity and activism at home, and helped lead to various policy reforms which gave tenants fairer rents, guaranteed tenure and compensation for land improvements. In 1903 the Land Purchase Act required landlords to sell to tenants if 75% of them wanted to purchase. Credit for purchase was made available through the State. In the period 1903 to 1920 9 million hectares had been transferred to ex-tenants. Under the new law, the previous owners were unable to retain ownership of the trees on land they were forced to sell. Realising that they had little chance of compensation, landowners cut large areas of their woodlands and sold the timber to the sawmillers. The new landowners needed annual revenue from annual crops, and had little incentive to preserve the trees.

Table 4	CHANGES IN FOREST AREA OVER TIME: EXAMPLE FROM FRANCE[42]		
Date	**Wooded Area as % Total**	**Date**	**Wooded Area as % Total**
3000 BC	80	1912	19
0	50	1963	21
1400 AD	33	1970	23
1650	25	1977	24
1798	14	1980	25
1862	17		

National Forest Services were also established during this period. The Greek Forest Service was created in 1836, and was originally concerned with identifying productive forest areas, and bringing them under management. This involved introducing silvicultural management to restore forest areas degraded through overcutting and grazing; reforestation of bare hills and in the vicinity of towns, and the opening of virgin forests.[38] Forest administrations were not always very popular with the rural populations. The early forestry professional generally represented urban interests, and perceived rural populations as a main cause of forest degradation. In Switzerland, in remote areas where economic change came slowly, foresters ran the risk of arousing the animosity of the local population: they were occasionally the targets of gunfire, and in one case the intended victims of a bombing. In the course of the 20th century, however, they were increasingly accepted in their role as advisers. The social gap between rural and urban areas had narrowed, and forest management assumed an important place in both the local and the national economy.[39]

NEW FOREST LAWS

The sweeping socio-economic transformations throughout Europe during this period also made possible the passage and subsequent enforcement of various Forest Laws to stem forest degradation. In Belgium a law was passed in 1847 concerned with deforestation and land clearance, and the state began to encourage reforestation by providing subsidies to communities and individuals for tree planting. In the Netherlands, the Forest Law of 1922 obliged all forest owners to replant clear cut areas within three years with timber trees. It also granted tax compensation on forestry generated income. Local communities (municipalities) were granted interest free loans for afforestation of their uncultivated lands. The notion that forests can be saved by forest laws alone, irrespective of social and economic contexts, is a myth which has frequently distorted views of present-day problems in developing countries.[40]

EXPANSION OF FOREST RESOURCES IN THE 20TH CENTURY

The 20th century was characterised by an expansion of forest resources throughout Europe, including increases in forest area, standing volumes per hectare and annual increments. The expansion has been widely encouraged throughout Europe through the adoption of official reforestation policies, technological advances and financial incentives since the turn of the century – as mentioned above. In many European countries the goal of rebuilding forests has been largely achieved. Nordic countries, such as Finland, Norway and Sweden, have raised their standing volume of wood in the 20th century. And in mountainous countries such as Switzerland, timber reserves in the last one hundred years have accumulated to the extent that they are now sufficient to meet most of the domestic demand for wood.[41]

Box 6   THE GRADUAL LOSS OF COMMUNITY RIGHTS IN DREVDAGEN, SWEDEN[46]

Drevdagen is a small village of 140 inhabitants in a remote area of northwest Sweden. The forest surrounding the village was originally owned by the settlers of the area who can be traced back to the 1300s.The forest has played an important role in the traditional livelihoods of villagers as a source of firewood, timber for household use, medicinal plants, an area for hunting and grazing animals. Lichen *(Cladonia rangiferia)*, which grows on the forest floor, was used as a winter fodder for mountain cows. Title deeds to the forest were based on ratified boundaries in 1751 and formed the legal basis for taxation until the end of the 1800s. In the late 1880s the forests grew in commercial value as a source of sawn timber, and forest ownership was transferred from the villagers to the government. Some residents of Drevdagen claim that the transfer of property rights was an illegal confiscation of private land, because the Swedish constitution requires such decisions to be made in court. In the 1960s Domänverket, the forest-owner parastatal, introduced large scale technology which reduced demand for labour, resulting in loss of local jobs, outmigration and subsequent closing of the village shop and post office. In their struggle to survive, the community has produced rural development proposals based on the sustainable development of the forests and natural resources. Experience from Drevdagan illustrates the gradual erosion of community rights over the forest since the 1860s.

1865: Loss of rights to harvest and sell timber
1894: Loss of rights to harvest timber for household needs
1950: Permission required to collect firewood
1966: Loss of rights to hunt moose
1973: Loss of fishing rights
1980: Fuelwood collection restricted to logged areas; and
      Loss of rights to harvest lichens for fodder

Although France has lost most of its natural forests, area under trees has almost doubled since the early 19th century. The steady increase in area under forest from the 1800s can be related to the equally steady decline in the proportion of French population dependent on agriculture. The French National Forest Fund established in 1946 financed some two million hectares of afforestation.[43]

INTENSIFICATION AND ITS EFFECTS ON COMMUNITIES

Driven by global market competition, most European countries with large state and private forest enterprises adopted industrial wood-production and processing techniques and practices in the 20th century. These included the mechanisation of planting and harvesting, drainage, use of chemical fertilizers and pesticides, use of fast growing non-indigenous tree species planted in monocultures; large scale production; increasing efficiency of raw material utilisation and manufacturing. Some Nordic countries have also deliberately developed new products and technologies which are less affected by global competition – such as short rotation wood production, high quality paper; and reorganised forest industries through company fusions to rationalise production and increase ability to invest.[44] The intensification process has greatly increased forest productivity across Europe, which has steadily grown since the 1950s. European wood production has increased by 18% since 1965[45] and has generated a timber surplus in some areas, although with social and environmental costs.

In some areas small forest owners and rural and urban communities have successfully adapted to technological developments, and are actively involved in the wood production, processing and marketing. This has generated incomes, employment and other significant social benefits. This is certainly not always the case. The intensification and expansion of commercial forestry has often been at the expense of local communities, biodiversity and landscapes. Rural people have paid high costs associated with the commercial exploitation of timber including loss of jobs and access to resources. Large-scale public and private enterprises rarely provide local communities with any economic benefits beyond employment. Indeed, intensification of wood production and processing has contributed to rural-urban migration across the region. In some areas local labour has been replaced by low-cost migrant work forces. Profits are rarely reinvested at source or into sustainable practices, and there can be serious conflicts of interest involving forest agencies, timber companies, local communities and environmental groups at a local level, and further afield. Some communities are challenging the local costs of commercial timber exploitation, and fighting to reclaim forests for local benefits, see below.

STAGNATING TIMBER REVENUES IN CENTRAL EUROPE

It is now increasingly difficult to harvest and market timber throughout central Europe. All European countries are more dependent on trends in the global trade in timber and timber products than before. Costs of harvesting and transporting timber have risen, while revenues earned from timber have stagnated. Labour costs have risen considerably since the 1950s. Declining energy prices have reduced transport costs, allowing timber that is exploited cheaply (in the tropics, Eastern Europe, and Canada) to penetrate European markets. This also applies to timber imported from Scandinavia, where costs are reduced by large scale, fully mechanised logging operations. Wood also has to face stiff competition from other materials. The steel industry has been heavily subsidised for decades in attempts to preserve jobs, and the inexpensive energy used to process both steel and concrete constitute a form of indirect subsidies.[47]

With the exception of the Nordic countries, profitability in the forest sector is poor within western Europe.[48] The contribution of the forest sector to Gross National Product (GNP) varies greatly between states. In the 1990s, it accounted for between 7-8% in Finland ; between 4-5% in Sweden and Norway, 1-2% in France and Germany, and less than 1% in the United Kingdom. Annual removals of wood per hectare are only about two thirds of the annual increment, resulting in a gradual build up and aging of the growing stock throughout Europe.

CHANGING CONSUMPTION

Patterns of wood consumption have changed considerably since the middle of the 20th century, with an increase in demand for sawnwood and panel products used in the construction and furniture industries; paper and board for use in packaging, printing and writing. The consumption of these products has increased by about 30% since the middle of the 20th century. Much of this demand has been met by intensifying production within Europe, particularly in Nordic countries, although cheap imports of wood products have increased from Eastern Europe. Nordic regions produce 45% of pulpwood, while Western countries produce about 36% of the industrial roundwood supply.[49] Since the 1960s there have been growing demands for more recreational and leisure facilities in forests throughout Europe. Forests are still important for providing a range of non-timber products, firewood and charcoal for use in household heating and cooking, resins, berries, mushrooms, honey, nuts for local and commercial use, although most people do not rely on these for their livelihoods. While the consumption of fuelwood is still important, it is no longer a major forest product in western Europe.

MULTIPLE USE AND OPEN INSTITUTIONS IN THE LATE 20TH CENTURY

Forest policy and the objectives of forest management have changed considerably across Europe in the late 20th century, with nearly all countries now committed to sustainable, multiple-use forest management policies. These take into account the forests' recreational, landscape, protective and conservation roles, as well as its productive ones. The multiple use concept partly reflects wider changes in the values of urban and industrial European society.

RURAL – URBAN TRANSITIONS

The population of western Europe has become increasingly urbanised during the last century, as a result of loss of employment opportunities caused by the intensification, concentration and mechanisation of agriculture and forestry, and demand for labour in urban areas. Although rural areas account for more than 80% of Europe's landmass, only 25% of the total population live and work in such areas. The percentage of the population employed in the primary sector (farming, fishing, mining and forestry) is now less than 5%.[50] For example, the reduction from 63,000 workers in forestry in Finland in 1982, to only 28,000 in 1993 was largely due to an increase in mechanised harvesting.[51] Although the decline has leveled off in the northern and central parts of Europe in recent years,[52] outmigration is still a problem in remote mountainous areas across the region.[53] In some areas, new trends in ecotourism, leisure activities, second homes and homes for retired people in rural areas are providing new employment opportunities, although rural incomes are markedly lower than urban ones, particularly in the south.[54]

Urbanisation has a number of implications for community involvement in forest management. Firstly, the outdoor leisure activities of urban communities have generated new demands on rural forests, and forest managers are having to adapt existing forests to accommodate new recreational interests.[55] In the Netherlands, for example, recreation within forests became an important issue in the second half of the 20th century. In the late 1990s it was estimated that 200 million individual visits are paid to the forest annually.[56] Many new urban forests and green spaces have been created throughout Europe.[57] The participation of urban communities in rural and urban forest management raises particular questions, approaches and technical considerations because of their distance from resources and diverse interests of user groups.[58] Secondly, urbanisation has undermined many of the traditional rural social institutions that once played a key role in organising rural life and resource use. Concern over the declining rural economy has generated EU interest in the role forestry can play in rural development in marginal areas. European Governments are committed to maintaining employment and standard of living in rural areas, and are seeking viable rural development activities particularly in economically marginal areas. The Cork Conference in 1996 agreed that an integrated set of measures were required to support rural communities, and subsequent proposals for the Ministerial Conferences on the Protection of Forests in Europe have supported the role of forests in rural development.[59] Rural communities clearly have a role to play in taking up forest-based opportunities supported through EU funds.

GROWING ENVIRONMENTAL AWARENESS

The adoption of sustainable, multiple use forest policies also responds to the growing understanding of the role forests play in maintaining biodiversity and providing environmental benefits.[60] Exotic tree species and intensive silvicultural systems have had a signficant impact on natural ecosystems within western Europe, and have led to concerns that despite the increase in quantity, there has been an overall decline in forest quality in temperate and boreal regions. New insights into ecological processes and principles, changing public demands and the influence of global economic factors affecting forest management have had important implications for the foresters' role, and organisational design,[61] to be discussed shortly.

Political awareness of ecological and environmental issues in Europe has grown considerably over the past forty years, particularly since UNCED in 1992, and the Pan European Ministerial Conferences on the Protection of Forests in Europe.[62] Many European countries are attempting to improve the conservation values of forests and to enhance biodiversity without jeopardising the economics of wood production. International and national environmental NGOs, such as the WWF and Greenpeace, also have considerable influence on National forestry debates, often directly lobbying public and private forest agencies to change their silvicultural practices towards greater use of broadleaf species, and protection of old-growth areas.

Some industries have also responded to public pressure by paying more attention to the origins of raw materials and improving their production processes. Nordic forest industries have made progress towards cutting down harmful emissions from their mills. Consumers and retail outlets of forest products are another powerful force guiding forest practice. Buyers are increasingly demanding assurance that forestry practices are environmentally sound, lending support to various ecolabelling and timber certification schemes, such as the Forest Stewardship Council (FSC) and Pan-European Forest Certification (PEFC), and supporting paper recycling projects.[63]

Many new community based forest initiatives have been motivated by environmental concerns. Several of the case studies within this profile testify that communities have a unique capacity to integrate environmental, social and economic objectives management, clearly providing opportunities for environmentalists to work with communities for the benefit of biodiversity conservation outside protected areas. In general, despite the increase in environmental awareness throughout the region during the last twenty years, significant ecological problems and threats persist, such as the logging of old growth forests in parts of Europe, and consumer indifference to environmental problems. Some of these problems are discussed in more detail in Part III.

Changing Democracy – Changing Forest Agencies

Europe has a long tradition of representative democracy based on elections, referendums, legal appeals, etc. However, within the last 15-20 years a new political wave within civil society, representing diverse perceptions, values and needs has been challenging the legitimacy of centralised and hierarchical management institutions everywhere, including those involved with forest management.[64] Giddens (1999)[65] has coined the phrase *'democratising democracy'* to characterise the demand for more consultation and involvement from civil society in decision making all the year round – not just at election time, and for more transparency and accountability within public institutions.

Growing public scepticism towards all centralised management agencies throughout the western world contrasts sharply with public faith in professionals in the 19th and early 20th centuries.[66] Shaped by prevailing social values at the turn of the century, forest agencies often behaved like a scientific and technological aristocracy serving the dominant commercial paradigm. Today, state forest policies and practices are being widely questioned and forestry professionals and organisations everywhere are being challenged to adapt their assumptions, values, attitudes, institutional structures and processes to accommodate the diversity and complexity of the interdependence of eco- and human systems. In practice, foresters are being asked to manage more open and democratic processes of public involvement in forest management.[67] Forest agency change typically involves the development of new policies; staff retraining; new programmes and procedures to enhance the participation of interest groups; new organisational arrangements and partnerships, such as the merging of forestry and environmental agencies, and employment of social scientists.[68] Many of these changes are providing new opportunities for different stakeholder groups in forest management, including rural groups.

Greater CIFM is closely related to the more general processes promoting pluralism and democracy within society, and changes within legislation and management institutions. Whilst uneven, some aspects of Western society do seem to be characterised by a growing pluralism and confidence at the grass roots, which may be encouraging some communities to get more involved in forestry initiatives. However the scope of these movements – and the changes within public forest institutions – should not be overstated. Traditional power relations and deeply embedded informal rules can work to thwart wider participation, openness and transparency of decision making.

CONCLUSION: FOREST HISTORY AND COMMUNITIES

The history of people-forest relations in Europe reveals trends familiar to many regions. Forests have provided essential resources for many of the major transformations in social history, but at a cost to natural ecosystems and to poorer rural communities. Through their conversion, they provided a source of fertile land which facilitated the advent of settled agriculture from the Neolithic era onwards. They were key to the expansion of the Roman Empire 2000 years ago, and to establishment of Kingdoms throughout Medieval Europe. They provided crucial sources of fuelwood and charcoal underpinning a period of early industrialisation, trade and colonial expansion in the 17th and 18th centuries. They have long continued to provide essential goods and services, such as fuelwood, fodder, and food, to marginalised rural peoples with limited resources. However, as history frequently reveals, where there is big money to be made from timber and forest products, regulation of forest resources has been difficult to achieve. Until the substitution of charcoal and fuelwood for oil and gas in the late 19th century, Europe's forests were almost uncontrollably depleted, except in the less densely populated areas, bequeathing a largely deforested landscape from which mainland Europe has been struggling to recover ever since.

The history of CIFM in Europe is embedded in this history of political and economic change. There are many examples of community ownership, rights and responsibilities in early European forest management. While it is important not to romanticise the nature of these institutions, it is evident that they provided access to resources for many rural peoples. These institutions were gradually undermined as they came into conflict with newly emerging state and private interests. As the forces of modernisation swept through Europe from the 17th century, the Age of Enlightenment – and its associated notions of progress, liberal reform, and economic expansion – had a dramatic influence on the way resources were used and organised. A pervasive discourse from the 1700s was to see the agricultural practices of rural people as the main threat to forest resources, and the commons as an obstacle to progress. The appropriation and privatisation of the commons by newly emerging states, and the liquidation of customary rights constituted a major disruption of traditional forms of CIFM in Europe, often benefiting the already socially privileged, and frequently resulting in considerable hardships for poorer rural people. Some community based institutions were replaced by newly formed commune or municipal forests, which were frequently resisted because they took direct decision making away from villagers. In other areas, loss of community involvement was offset by a new dependency on employment in industries and the rise of the modern welfare state which provided material benefits in exchange for loss of local control. At the start of the new millennium, the tide has turned once more, and there are indications of new interest and new actions for CIFM.

Notes

[1] Harrison, R.P. (1992): *Forests: the Shadow of Civilisation*. Chicago: University of Chicago Press.

[2] Schama, S. (1995): *Landscape and Memory*. London: Harper Collins.

[3] Rietbergen, S. (2001): "History and Impact of Forest Management", in Evans, J. (Ed) (2001): *The Forests Handbook*,Vol.2. Oxford: Blackwell Science Ltd.

[4] Rackham, O. (1986): *The History of the Countryside. The Classic History of Britain's Landscape, Flora and Fauna*. London: Weidenfeld & Nicolson.

[5] Delcourt, P.A. & Delcourt, H.R. (1987): "Long Term Forest Dynamics of the Temperate Zone". *Ecological Studies* 63.

[6] Rietbergen, S. (2001) op.cit.

[7] Rackham, O.(1986):op.cit.

[8] Tuite, P. & Brown, D. (1998): "Ireland" in Shepherd, G; Brown, D; Richards, M; & Schreckenberg, K. (Eds) (1998): *The EU Tropical Forestry Sourcebook*. London: Overseas Development Institute, Brussels: European Commission.

[9] Meiggs, R. (1989): *Farm Forestry in the Ancient Mediterranean*. ODI Social Forestry Network Paper 8b. London: Overseas Development Institute.

[10] Navone, P. & Sheperd, G. (1998): "Italy" in Shepherd, G; et al. (Eds) (1998): op.cit.

[11] Navone, P. & Shepherd, G. (1998): op.cit.

[12] Rackham, O. (1986) p.65: op.cit.

[13] Shoard, M. (1987): *This Land is our Land. The Struggle for Britain's Countryside*. London: Paladin.

[14] Pasmore, H. (1981) & Drummond, M. (1981), in Forestry Commission (1981): *Explore the New Forest*. London: Her Majesty's Stationary Office. Heathcote, T. (1997): *Discovering the New Forest*. Tiverton: Halsgrove.

[15] Rackham, O. (1995): *Hedges and Hedgerow Trees in Britain: A Thousand Years of Agroforestry*. ODI Social Forestry Network Paper 8c. London: Overseas Development Institute. Also Rackham, O. (1986): op.cit; Shoard, M. (1987): op.cit.

[16] Sandberg, A. (1998): "Against the Wind: On Reintroducing Commons Law in Northern Norway". *Mountain Research and Development* 18 (1): 95-106.

[17] Morin, G-A. (1996): "France", in Morin, G-A; Kuusela, K; Henderson-Howat, D.B; Efstathiadis, N.S; Oroszi, S; Sipkens, H; Hofsten, E,v, & MacCleery, D.W. (1996): *Long Term Historical Changes in the Forest Resource*. Geneva Timber and Forest Study Papers, No.10. New York and Geneva: United Nations.

[18] Tuite, P. & Brown, D. (1998): op-cit.

[19] Arnold, M. & Calender, T. (1998): "Sweden". In Shepherd, G. et al. (Eds) (1998): op.cit.

[20] Veron, P. Federspiel, M. & Shepherd, G. (1998): "Belgium". In Shepherd, G. et al. (Eds) (1998): op.cit.

[21] Navone, P. & Shepherd, G. (1998): op.cit.

[22] Küchli, C. (1997): *Forests of Hope. Stories of Regeneration*. London: Earthscan.

[23] Sipkens, H. (1996): "The Netherlands", in Morin, G-A; et al. (1996):op.cit.

[24] Küchli, C. (1997): op.cit.

[25] Küchli, C. (1997): op.cit.

[26] Groome, H. & Richards, M. (1998): "Spain", in Shepherd, G; et al. (Eds) (1998): op.cit. See also Brouwer, R. (1995): *Planting Power: The Afforestation of the Commons and State Formation in Portugal*. Proefschrift Landbouwuniversiteit Wageningen.

[27] Campbell, B.M.C. & Godoy, R.A. (1992): "Commonfield Agriculture: the Andes and Medieval England Compared". In Bromley, D.W. (Ed) (1992): *Making the Commons Work: Theory, Practice and Policy*. USA: Institute of Contemporary Studies.

[28] McGranahan, G. (1991): "Fuelwood, Subsistence Foraging, and the Decline of Common Property". *World Development* 19 (10): 1275-1287.

[29] Sandberg, A. (1998): op.cit.

[30] Küchli, C.(1997): op.cit.

[31] Küchli, C.(1997): op.cit.

[32] Tuite, P. & Brown, D. (1998): "Ireland", in Shepherd, G. et al. (Eds) (1998): op.cit.

[33] Küchli, C. (1997): op.cit.

[34] Kuusela, K. (1994): *Forest Resources in Europe 1950-1990*. European Forest Institute Research Report 1. Cambridge, UK: Cambridge University Press.

[35] Farmers in the southern part of the Black Forest had successfully grown Dutch firs for centuries. Küchli, C. (1997): op.cit.

[36] Bedel, J. & Brown, D. (1998): "France", in Shepherd, G. et al. (Eds) (1998): op.cit.

[37] Sipkens, H. (1996): op.cit.

[38] Varelides, C. & Richards, M (1998): "Greece", in Shepherd, G, (Eds) (1998): op.cit.

[39] Küchli, C. (1997): op.cit.

[40] ibid.

[41] Stanners, D. & Bourdeau, P. (1995): *Europe's Environment. The Dobris Assessment*. European Environment Agency: Copenhagen.

[42] Source: Mather, A.S. (1990): *Global Forest Resources*. World Resources Institute: World Resources 1992/93.

[43] Morin, G-A. (1996): op.cit..

[44] Hytönen, M. & Blöndal, S. (1995): "Timber Production and the Forest Industry". In Hytönen, M. (Ed) (1995): *Multiple-Use Forestry in the Nordic Countries*. Helsinki: METLA.

[45] Stanners, D. & Bourdeau, P. (Eds) (1995): op.cit.

[46] Source: Halvarsson, H. (1998): "Local Forest Management: Hope for the Future in our Struggle for Survival". *Forests, Trees and People Newsletter* No.36/37: 34-40.

[47] Küchli, C. (1997): op.cit.

[48] Stanners, D. & Bourdeau, P. (Eds) (1995): op.cit.

[49] Stanners, D. & Bourdeau, P. (Eds) (1995): op.cit.

[50] FAO/ECE/ILO (2000): *Public Participation in Forestry in Europe and North America*. Working Paper 163. Geneva: International Labour Office.

[51] Hytönen, M. & Blöndal, S. (1995): op.cit.

[52] Glueck, P. (1997): *Sustainable Forestry in the Context of Rural Development*. In FAO/ECE/ILO (1997) op.cit.

[53] Butt, N. & Price, M.F. (2000): *Mountain People, Forests and Trees: Strategies for Balancing Local Management and Outside Interests*. Synthesis of an Electronic Conference of the Mountain Forum April 12- May 14, 1999. Harrisonburg: Mountain Forum.

[54] Glueck. P. (1997) op.cit.

[55] Fritzboger, B; & Sondergaard, P. (1995):"A Short History of Forest Uses". In Hytönen, M. (Ed) (1995): *Multiple-Use Forestry in the Nordic Countries*. Helsinki: METLA.

[56] Sipkens, H. (1996): op.cit.

[57] Jensen, F.S. (1995): "Forest Recreation". In Hytönen, M. (Ed) (1995): *Multiple-Use Forestry in the Nordic Countries*. Helsinki: METLA.

[58] FAO/ECE/ILO (2000): op.cit.

[59] Glueck, P. (1997): op.cit. Also Ministerial Conference on the Protection of Forests in Europe (2000): *The Role of Forests and Forestry in Rural Development – Implications for Forest Policy*. International Seminar, 5-7 July, Vienna, Austria.

[60] Dudley, N. (1992): *Forests in Trouble. A Review of the Status of Temperate Forests Worldwide*. WWF International, Gland, Switzerland.

[61] Kennedy, J.J; Dombeck, M.P. & Kock, N.E.(1998):"Values, Beliefs and Management of Public Forests in the Western World at the Close of the Twentieth Century". *Unasylva* 49 (192): 16-26.

[62] Strasbourg (1990); Helsinki (1992); Lisbon (1998).

[63] Hytönen & Blöndal (1995) op.cit.

[64] Anderson, J; Clement, J; & Crowder, L.V. (1998):"Accomodating Conflicting Interests in Forestry – Concepts Emerging from Pluralism". *Unasylva* 49 (194): 3-10.

[65] Giddens, A. (1999): *Runaway World. How Globalisation is Reshaping our Lives.* London: Profile Books.

[66] Anderson et al. (1998) op.cit. Also, Ockerman, A. (1999): "Changing Forests, Changing Ideas: A Cultural History and Possible Futures for Forestry". In *Boreal Forests of the World. Integrating Cultural Values into Local and Global Forest Protection. Proceedings of the 4th Biannual Conference of the Taiga Rescue Network, October 5-10, 1998, Tartu, Estonia.* Tartu: Estonian Green Movement.

[67] These changes do not preclude the value of technical knowledge, but rather expand professional knowledge and skills to serve a wider range of social values.

[68] Hytönen, M. (1995): "History, evolution and significance of the multiple-use concept". In Hytönen, M. (Ed) (1995): *Multiple-Use Forestry in the Nordic Countries.* Helsinki: METLA.

Figure 3

MAJOR FOREST TYPES IN EUROPE

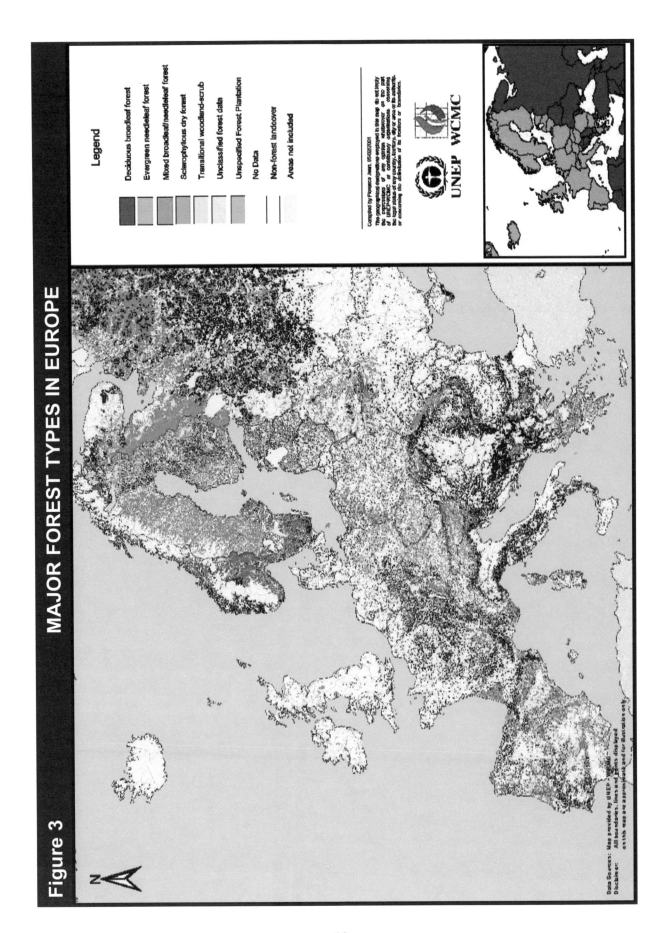

Legend

- Deciduous broadleaf forest
- Evergreen needleleaf forest
- Mixed broadleaf/needleleaf forest
- Sclerophyllous dry forest
- Transitional woodland-scrub
- Unclassified forest data
- Unspecified Forest Plantation
- No Data
- Non-forest landcover
- Areas not included

Compiled by Florence Jean, 05/02/2001

The geographical designations employed in this map do not imply the expression of any opinion whatsoever on the part of UNEP/WCMC or contributory organisations concerning the legal status of any country, territory, city or area or its authorities, or concerning the delimitation of its frontiers or boundaries.

UNEP WCMC

Data Sources: Map provided by UNEP-WCMC
Disclaimer: All boundaries, lines and points displayed on this map are approximate and for illustration only.

32

Part III

FOREST BIO-REGIONS IN EUROPE

NATURAL ECOSYSTEMS

The natural forests of Europe are usually divided into three broad types related to the major climatic regions, which include:[1]

◆ Boreal coniferous and mixed broadleaf forests

◆ Temperate deciduous broadleaf forests

◆ Mediterranean semi-arid, evergreen, sclerophyllous forests

See Figure 3 for major forest types in Europe. Most of Europe's natural vegetation has been modified by human activities during the last 10,000 years, and forest management practices during the last centuries have resulted in substantial changes to the remaining forests, in terms of species composition and structure. Mature and natural forest ecosystems are very rare in Europe. It is estimated that only 2-3 per cent of the forest estate in Western Europe can be considered as relatively intact natural forest.[2] The main habitat types, principal changes, and ecological concerns are identified briefly below.[3]

THE NEEDLELEAF AND MIXED FORESTS OF NORTHERN AND BOREAL REGIONS

In the northern hemisphere, boreal woodlands form a continuous belt around the whole Earth, covering much of Scandinavia and Northern Russia with trees such as Norway spruce *(Picea abies)*, Scots pine *(Pinus sylvestris)* and downy birch *(Betula pubescens)*. The summer temperature in the boreal region ranges from 10 to 15 degrees centigrade. In the winter months the mean temperature varies between -3 to +2 degrees centigrade. In the Northeast, winter temperatures of -20 to -40 degrees centigrade, with desiccating winds can damage needles and be fatal to trees on the Northeast boundary of the region. The precipitation varies from 400 mm in the north to 800 mm in the south, and it can be as high as 2000 mm on the west coast. Much of the precipitation falls as snow, which can last from 4 to 7 months. Glacial and alluvial sands, gravels and moraines are the most frequent soil parent materials in the boreal zone. Boreal trees are adapted to a fire ecology and for the invasion of treeless sites. Fires start by lightening and are usually repeated on average once every 50 years on dry sandy sites, to 150 years on moister sites. Boreal forests are very poor in vascular plants, and numbers rarely exceed 250 species. They are, however, rich in mosses and lichens. The boreal region is the only forest ecosystem in Europe where large numbers of carnivores such as wolves, wolverine, lynx and bear, can be found in extensive intact habitats. However, in western parts the wolf and the brown bear have become very rare.

In the Nordic region there has been an important shift towards the industrialisation or intensification of forest management for wood production. This has involved the mechanisation of the planting and harvesting of fast-growing conifers but also of broad-leaved species such as

PLATE 2: Boreal coniferous and mixed broadleaf forests, Finland
(Photo: WWF-Canon/Mauri Rautkari)

birch (*Betula* spp.), planted in monocultural crops, to be clear felled at regular intervals, with regular applications of fertilizer. Intensive silvicultural practices tend to be limited to areas of forest owned by large forestry companies or to publically owned forests, with private forest owners practising more mixed and less intensive systems.

THE DECIDUOUS BROADLEAF FORESTS OF TEMPERATE REGIONS

The Atlantic region was once dominated by broad-leaved deciduous forests of oak (*Quercus* spp.) and beech *(Fagus sylvatica)*, but silvicultural management has brought about large changes in this region, with a trend towards coniferous trees. This area is subject to maritime influences from the west, and continental influences in the east. Summers are relatively warm, with the mean temperature in July varying between 15 degrees centigrade in the Northwest to more than 20

degrees centigrade in the Southeast. The winters are mild lasting between 2 to 4 months. In January the mean temperature varies from +5 degrees centigrade in the west to -5 degrees centigrade in the east. Much of the precipitation falls in the summer months, which varies from 1000 mm on the Atlantic coast to 600mm in the east. Soil parent materials are more rich and varied than in the boreal zone, representing layers from many geological periods. Winds are often strong, especially in the western Atlantic part of the zone and can cause damage in forests. The extensive planting of non-native species such as Sitka spruce *(Picea sitchensis)* has resulted in the decline of the original, ecologically much richer composition of herbaceous plants and animals.

In western and central regions there has been an extensive replacement of mixed forests by coniferous plantations, especially in German speaking countries in the latter part of the 19th

34

century. Since the last decade of the 20th century there has been a more cautious application of modern forestry methods in some countries, combined with a recognition of the need to for greater environmental concern, and a general trend towards more mixed forests. In other countries, such as France and Germany, intensive wood production practices are still common.

<hr>

Box 7   **COMMUNITIES AND NON-TIMBER FOREST PRODUCTS IN THE EUROPEAN BOREAL ZONE**[4]

The boreal forests provide a wide range of non-timber forest products (NTFPs) such as berries, mushrooms, ornamental foliage and lichens and medicinal plants. NTFPs can also include fish, wild game, reindeer grazing, as well as the forest's recreational and watershed protection values, and the forest's role as a carbon sink. The diversity of NTFP species and use is vast. It is estimated that Finland has 50 indigenous berry species, of which 37 are considered edible and 16 picked for human consumption.[5] The following list represents a fraction of the NTFPs of medicinal, nutritional, cultural, and commercial value available from the European boreal forests:

Lingonberry	*Vaccinium vitus-idea*
Bilberry	*Vaccinium myrtillus*
Cloudberry	*Rubus chamaemorus*
Reindeer Lichen	*Cladonia alpestris*
Sphagnum moss	*Sphagnum* spp.
King bolete mushroom	*Boletus edulis*
Norway Spruce	*Picea abies*

NTFPs provide market and non-market benefits for households, communities, and enterprises throughout the boreal region; although in post-industrial societies in Northern Europe, the average citizen is not dependent on NTFP resources for subsistence. In the Nordic countries there is very little restriction of movement on NTFP harvesters. The "every-man's right" recognized in these countries allow free movement on almost all land even if it is privately owned. Despite its value, the incentives to participate in the commercial NTFP harvest has declined in recent years.

Despite a number of problems to overcome, the development of NTFPs offers great opportunities for communities and enterprises in the forests in the European boreal region, and the potential to contribute to diverse, sustainable, multiple-use forests. European boreal countries have experienced increased unemployment over the last thirty years due to mechanization, restructuring of industries, and declining resource stocks. Forest-based communities and workers have been left with diminished economic means, and are struggling to regain healthy economies in tune with healthy environments. Many communities are looking for local solutions. The current non-formal nature of much of the NTFP activity suits households and communities.

The management of NTFPs presents complex challenges. There is a need to broaden the view of the forest's economic, ecological and social values to include NTFPs. A greater flow of information is required about the ecology, economy and social issues surrounding NTFP development and management. Adaptive management planning, non-linear thinking, and the involvement of multiple stakeholders are required. Regulatory systems for NTFP management need to be designed so that they protect NTFP ecology but do not result in unnecessarily reduced access for small enterprises and individual users. Some regulatory systems – involving difficult bureaucratic processes and fee based permits – may make it difficult for small actors with less capital to participate in the commercial and recreational harvest. Another challenge for small enterprises is locating appropriate small scale processing equipment.

PLATE 3: Deciduous broadleaf forest, Switzerland.
(Photo: WWF-Canon/Michèle Dépraz)

SCLEROPHYLLOUS DRY FOREST OF THE MEDITERRANEAN REGIONS

Mixed evergreen forests occur around the Mediterranean and Black Sea. In these regions the summers are warm or hot, and the winters are mild. The mean temperature in the warmest months ranges from +10 to 35 degrees centigrade, and that of the coldest months from +2 to 18 degrees centigrade. Precipitation falls mostly in the winter, and in general ranges between 500 to 800 mm, although in parts of inland Spain and Turkey it can be as low as 200 mm. Treeless steppe and desert are the natural vegetation types in the driest areas. Snow falls regularly in the mountains and northern parts of the zone. Around the Mediterranean Sea; inland Spain; Southern Greece and Italy the main natural vegetation type is sclerophyllus scrub with some woodland, dominated by evergreen oaks: holm oak *(Quercus ilex; Q. pubescens)* kermes oak *(Q. coccifera)*, cork oak *(Q. suber)* which is widely planted for its valuable bark. In the eastern part of the zone, oaks are frequently accompanied by pines, especially Aleppo pine *(Pinus halepensis)*; Calabrian pine *(P. brutia)*. The maritime pine *(P. pinaster)*, and stone pine *(P. pinea)* grow in coastal regions. At higher altitudes, black pine *(P. nigra)* is common. Above 2000 metres junipers *(Juniperus* spp.) are common and may form woodland as in parts of central Spain. *Cupressus sempervirens* and *Cedrus brevifolia* are found in parts of Greece and Cyprus.

The southern region of Europe, with the exception of some areas in Spain and Portugal, has witnessed the abandonment of many traditional forestry practices, such as coppicing for firewood and the collection of barks, resins, acorns and tannin, often as a consequence of rural depopulation. The region has been unable to compete with cheap wood imports from highly productive Nordic and eastern countries, and the area has seen a decline in the management of its forests for wood production. Many forests have become neglected or abandoned, increasing the risk of fire, see Box 8 below. However, innovative uses of NTFPs can play important roles in rejuvenating

36

rural economies as well as contributing to the conservation of ecosystems in southern Europe, see Part V.

ALLUVIAL FORESTS

These occur in all climatic zones. They consist of two main types: hardwood forests with oak (*Quercus* spp.), alder (*Alnus* spp), ash *(Fraxinus excelsior)* or elm (*Ulmus* spp.); and softwood forests with willow (*Salix* spp.) and poplar (*Populus* spp.). Alluvial forests once fringed all European rivers. However many alluvial forests have disappeared since the flood plains of larger rivers have been used for human settlements. The few remnants along the Rhine and Danube represent important wildlife habitats.

MOUNTAIN FORESTS

In central European mountains, beech *(Fagus sylvatica)* is intermixed with fir (*Abies* spp.) and larch *(Larix spp.)* and followed by spruce (*Picea* spp.). At higher altitudes the mountain vegetation is dominated by rocky grassland and heath and dwarf pines (*Pinus mugo, P. uncinata*).

SOME ECOLOGICAL CONCERNS

DECLINE OF FOREST QUALITY

While the area of forests has remained constant or even increased in the last half century,[6] conservation organisations are concerned that the statistics mask a major degradation in temperate forest quality. From an ecological point of view, one of the elements contributing to quality is the degree of forest *naturalness.*[7] While it is difficult to determine the composition of an original forest, and most forests in the region have been altered by human activity for hundreds if not thousands of years, there is concern that the fragmentation, isolation and loss of natural forest ecosystems, and replacement by plantations of non-native species,

PLATE 4: Sclerophyllous dry forest, Greece.
(Photo: WWF-Canon/Michèle Dépraz)

leads to a loss of biodiversity, simplification of ecosystems and lowered ecosystem resilience in the face of change. Increasing areas of forests are managed to produce low-grade timber materials for pulp and paper manufacturing, and plantations do not support the same range of species or ecological functions. The intensification of management associated with commercial forestry, such as clear felling, drainage, pesticide and fertilizer use, heavy machinery use, can have detrimental environmental impacts. Environmental groups argue that forests should be managed for a much wider range of goods and services than just timber, such as biodiversity conservation, watershed protection and recreation.[8]

TEMPERATE FORESTS AND POLLUTION

During the 1980s there was considerable concern that large areas of forest in Europe were showing symptoms of decline as a result of air pollution, including the effects of sulphur dioxide, nitrogen oxides and ozones. *Acid rain* was often used as a synonym for this atmospheric pollution. It was speculated that air pollution was damaging trees in areas, such as Scandinavia, which were far away from major pollution sources. Symptoms of decline include several features, depending on the species, including discolouration and cracking of leaves; premature leaf and needle fall; erratic branching of twigs; and loss of crown density; a general reduction in vitality and in some cases death. Several European-wide surveys were initiated to research tree condition and causes of forest die-back, which have provided a time series of large scale spatial observations, but they have not readily permitted cause and effect relationships to be identified.[9] In the transnational survey of 1992, 24% of European trees were considered to be damaged, that is, with a defoliation greater than 25%. The majority of critically affected conifer forests occur in Central Europe. Some broadleaf species such as *Fagus sylvatica* in Denmark and *Betula pubescens* in Sweden are also suffering from die-back. Improved data and analysis indicated that the original assumption that air pollutants were exclusively to blame for forest decline was too simplistic. The probable causes of discolouration and defoliation are a complex mix of adverse climatic conditions, diseases, nutrient

deficiencies, inappropriate management, and air pollution. Some countries such as former Czechoslovakia, Germany and Poland, consider air pollution to be a leading factor in the weakening of forest ecosystems. The largest and least debatable effects of air pollution are on forest biodiversity, particularly tree-living lichens, mosses and other sensitive plants.

BOREAL FORESTS AND CLIMATE CHANGE[10]

Of all the forest ecosystems, boreal forests are expected to be the most profoundly impacted by climate change, with potentially tremendous biological, social, and economic implications. However, there is considerable uncertainty about the scope and direction of changes. Some predictions estimate that the area covered by boreal forests may decrease by 25-40% due to increasing temperatures. Many of the present-day boreal forests could be replaced by temperate forest species, grasslands, and forest-steppe (in Russia). A total of up to two-thirds of the boreal forest is likely to undergo species and ecosystem changes. Winter temperatures may rise 4-5°C, with several models predicting an increase of 10 degrees or more. The hydrology of the boreal region is expected to change dramatically in some areas, reducing the availability of water, which could dry soils and increase fire risks. These forests would face unsuitable temperature, water, and soil conditions in their present range, and may be hindered from following more favourable conditions northward by natural impediments and barriers created by human land use patterns. However, given current technological limitations in climate modelling, uncertainties exist regarding the specific impacts, and scientists expect climatic "surprises" – unforeseeable consequences – as modern ratios of carbon dioxide accumulation exceed any known historic level. The predicted set of impacts of climate change on boreal forests has become an important issue for Scandinavian countries which own approximately 595,000 km^2 of the boreal forest.

FIRE HAZARDS

Although fire is an important element of the natural dynamics of the northern forests, fire is a

Figure 4   **FOREST FIRES IN SOUTHERN EUROPE 1989-1991**

Average Annual Number of Forest Fires, 1989-1991

- 250 or more
- 120-249
- 50-119
- 20-49
- 1-19
- 0

Data Sources: Average Annual Number of Forest Fires, 1989-1991 from 'The Dobrh Assessment', European Environment Agency 1995. ESRI Data and Maps -- World and Europe datasets.

Disclaimer: All boundaries, lines and points displayed on this map are approximate and for illustration only.

1000   0   1000 Kilometers

major cause of forest damage in Southern Europe, with countries such as Greece, Italy, Portugal, Spain and Turkey being particularly affected.[11] The frequency of fires and the vast areas affected each year, constitute a serious obstacle for investment in forestry and in forest management. In southern Europe a high population density and small scale forest ownership combine to increase the likely significance of a particular fire. The number of fires and their extent is very variable, depending on the climate and fire control policies. The frequency of fires in Spain and Portugal increased in the 1990s, with an annual average of 20,019 fires in Portugal and 17,429 fires in Spain. Greece had an annual average of 1,874 fires during the 1990s.

The causes of forest fires are similar all around the Mediterranean basin, and are generally beyond the control of forest policy. They include emigration; general collapse of traditional agricultural practices; non-complementary agricultural and forest activities in rural areas; urban development; chaotic development of coastal real estate projects; insurance crime and political protest. There has also been a growing tendency amongst the wider population to delegate the fighting of fire to professional firemen and other civil authorities, and to abandon traditional fire prevention activities. However, a number of countries are beginning to implement new programmes based on wider community involvement to help prevent and control fire, see Box 8 below.

PROTECTING FORESTS

Conservation organisations working within Europe are concerned that natural forest ecosystems are inadequately protected within the region. Many forest types have been reduced to tiny fragments, and large parts of remaining forests are species-poor plantations rather than forests in which natural ecosystem processes predominate.[12] Many forest types are either not represented in national protected area networks or are of insufficient size for proper protection, especially with respect to wildlife such as large predatory mammals and birds.

Finland still has good examples of virgin and ancient semi-natural forests. The largest proportion of these remnants is in the northern part of the country. The region along the Russian border (Karelia and Kainuu) still contains some very large old growth forests, which are considered especially important since they serve as a corridor for animal and plant species. The Kainuu region is currently under threat by the timber industry. The status of natural and semi-natural forests in the centre of Western Europe is the

In Portugal fire is a constant threat to forest land and seriously undermines the profitability of forestry. During 1998 more than 100,000 hectares of forest were burned as a result of 30,000 fires, occurring mainly during the summer. Work undertaken by the Forest Fire Investigation Brigade concluded that some 56% of fires were of human origin; 2% of natural origin and 42% of unknown origin. At least 29% of fires were started as a consequence of human negligence. Many fires are set on fallow agricultural land by shepherds to improve pasture productivity, but spread to forest stands, indicating the need for prevention strategies to be operational in and outside the forest area. Apart from being a constant feature in television news on summer evenings as well as a subject of political debate and of questioning government performance, forest fire has become an important issue in policy discussions because it seriously undermines the profitability of investment in forests. The fact that almost 90% of forests are privately owned with 83% under 3 hectares also makes it harder for forest owners to effectively commit to fire prevention. After years of relatively unsuccessful government action and investment focussing on fire fighting, new programmes have been adopted by the Portuguese government which encourage more participatory approaches. The National Plan for the Sustainable Development of Forests in Portugal (1998) aims to reduce the forest land burned by 20% between 1998-2003 and by 50% over the years 2003-2008, and supports the prevention and control of forest fires through establishing participatory teams of fire guards.

The new Forest Law (1996) establishes as a priority action the creation of voluntary Teams of Fire Guards *(Sapadores Florestais),* and deals with incentive and regulatory actions to involve local communities. This programme is a real partnership since neither the government nor forest owners individually have the means to implement the programme: the government lacks the human resources and the forest owners lack the funds. Teams of Fire Guards receive institutional and financial support from the government for professional training, purchase of equipment and operational costs. Private forest owners and community based associations are eligible to apply, as long as they are able to support 25% of the variable costs. The teams need to have at least 5 members, who receive 110 hours of training and a professional qualification. Fire prevention activities are restricted to local areas. One of the strengths of this project is the real involvement of local communities. Out of the 65 teams established by 2000, some 21 were from bodies governing common land forests. The training of team members also enhances the local capacity for sustainable forest management in general.

worst in the region. Some small patches exist in inaccessible regions in the Swiss and Austrian Alps. Belgium, Britain, Denmark, Ireland, Luxembourg and the Netherlands have little surviving natural forest. Despite small areas of natural woodland, and a long history of forest devastation, Greece contains one of the best remnants of ancient semi-natural forest in Europe in the Rhodopi-Paranesti region. It is home to significant populations of brown bear, wolf, and many birds of prey. The area is currently threatened by the construction of 3 dams on the Nestos river.

Concern about a decline in natural forests have generated a number of EU, National and NGO initiatives for an increase in protected areas, to protect biodiversity and related ecological, social and cultural values.[14] For example, the EU Habitats Directive came out of the Convention on Biodiversity adopted in Rio in 1992. This legislation establishes Natura 2000 – a network of areas designed to conserve rare, endangered species and habitats within EU member states. At the end of the 1990s over 200 projects had been funded with a total of over 134 million ECU. At an inter-ministerial level, two conferences on 'Environment for Europe', concerned with enhancement of biological and landscape diversity, have taken place in Sophia 1995 and Aarhus 1998. Much current work is on developing methodologies and indicators for monitoring and evaluating forest biodiversity in Europe.[15]

Despite these initiatives aimed at protecting forests in Europe, there are concerns that EU member states are jeopardising the last remnants of high value natural forests in Europe by failing to implement the EU Habitats Directive.[16] Some industry representatives have expressed concern about the amount of forest being removed from productive use and questioned the amount of land being put into protection.[17] There are currently 36 infringement procedures pending against member states for inadequate implementation. In Finland, where most of the forested land is owned by the Forest and Park Service (FPS) a row erupted when the FPS began logging old growth forests in north Karelia and southern Kainuu which were proposed for protection but not yet designated. The FPS were accused of deliberately logging the most important old growth forests areas proposed for protection under the EU Habitats directive. Problems have also occurred where private forest owners have logged areas proposed for protection to degrade their conservation values.[18] The high percentage of private forest ownership in Europe could influence the effectiveness of EU and governmental conservation and management of forest lands.

The WWF European Forest Scorecards, published in 2000, analysed how various European countries have implemented major international commitments to forest protection. Table 5 below indicates the scores on forest protected areas, out of a total maximum score of 100. Their analysis was based on scoring several factors including: data quality; government commitment; trends in protection; ecological representation; geographic distribution; management plans; quality of active management; and quality of protection. The fact that the highest score achieved was only 55 shows there is much room for improvement in all countries assessed.

WWF makes a number of recommendations to European governments and the private sector for forest protection in Europe:

1. Secure the protection of the remaining old-growth, relic and other high conservation value forests.
2. Use ecological criteria in the design and location of forest protected areas
3. Make the management of existing forest protected areas effective
4. Use a range of tools to create and manage forest protected areas in the future

However, the establishment and management of protected areas and national parks within Europe, often fails to adequately integrate social concerns and needs. Over 45% of national parks

Table 5	PERFORMANCE OF EUROPEAN COUNTRIES ON FOREST PROTECTED AREAS		
Country	Score on Forest Protected Areas	Country	Score on Forest Protected Areas
Belgium (Flanders)	55	Sweden	40
Finland	55	Belgium (Wollonia)	40
Spain	55	United Kingdom	40
Greece	53	Switzerland	39
Netherlands	50	Norway	34
France	48	Germany	25
Austria	45		

in Europe admit the existence of conflicts between park authorities and local communities over use restrictions. Only 39% claim to involve inhabitants in land use planning.[19] Clearly, efforts to ensure forest conservation need to cooperate more closely with rural populations to ensure that ecological and social aspects of sustainability are better integrated.

COMMUNITY INVOLVEMENT IN FOREST PROTECTION

Since the 1970s, many local communities and NGO activists have joined together to protect important patches of remnant and old growth forests of high conservation value throughout Europe.[20] In general, the significance of forests, woods and trees in Western Europe has shifted away from purely monetary and subsistence values. The public is placing increasing value upon the cultural, aesthetic, recreational and spiritual aspects of forests, woods and trees, often over and above their economic significance.[21] Woods are integral elements of local landscapes and are frequently important symbols of community identity, and sources of local and regional pride. To many people the 'tree' stands as a symbol for nature itself, which needs to be protected from the threat of urbanisation and over-development. Many important remnant woodlands and old growth forests within the region are threatened by urban sprawl and road-building which can have huge impacts on the quality of life of local people, as well as negative effects on biodiversity. Community involvement in forest protection in Europe takes many forms, from financial support to woodland conservation organisations and charities; local initiatives concerned with native woodland management and conservation, and direct action.

Community involvement may also take the form of protest and direct action. See Plate 5 depicting community resistance to the Newbury Bypass in the United Kingdom in the mid 1990s. Some new roads, which may take only a few minutes off travel time, have generated considerable community resistance because of the lack of public consultation and, in some cases, government avoidance of environmental impact assessments and EU legislation.[23] Direct action has variable success. It clearly raises the public profile of endangered woodland, and acts as an emotive communications and educational tool. Such publicity can encourage companies and governments to adopt more responsible policies and behaviour. For example, after the blockade of the ship carrying paper from old-growth forests in Finland to Holland (Plate 6), the Dutch paper publishers association issued a statement demanding a guarantee of old-growth free paper.[24] However, direct action and protest works most effectively when it is backed up by a solutions-

Box 9	CHARITIES DEDICATED TO WOODLAND PROTECTION: THE WOODLAND TRUST IN THE UNITED KINGDOM[22]

The Woodland Trust in the United Kingdom, which was established in 1972, aims to prevent further loss of ancient woodland; to improve woodland biodiversity; to increase the area of new native woodland, and to increase people's awareness and enjoyment in woodland. It has over 60,000 members who are committed to protecting and planting native woodland, and receives over 40% of its income (over £16 million in 1999) from the general public. It now owns 1,124 woodlands throughout the British Isles, covering some 18,200 hectares. Many of the woods they acquire are under threat from development pressure, leisure pursuits incompatible with conservation, or unsympathetic management. The Woodland Trust is the first major landowner in England and Wales to have all its woods certified under the FSC standard. It also has a community 'Woods on your Doorstep' scheme which supports community identification, fundraising, design and management of woodlands. This aims to create some 200 community woods for the new millennium.

PLATE 5: Tree Huggers of Europe. The site of the Newbury bypass in the UK.
(Photo: A. Testa)

PLATE 6: Blockade of the Finnish ship 1997. Activists protest about the use of Finnish old-growth forests for paper in Holland.
(Photo: X. Rimikes)

oriented approach; harnesses wider public support, and where local actors collaborate with other influential lobbyists.

SUMMARY

Europe contains a diversity of natural forest types: boreal coniferous forest in the north; broad-leaved deciduous forests in the temperate zone, and mixed evergreen forest in the Mediterranean area, all with many important local variations. Most forest ecosystems have been degraded by the intensification of forestry practices, and expansion of agricultural and urban areas. Only a tiny proportion of semi-natural forests survive. Forest ecosystems are threatened by intensive forest management practices; pollution; climate change and poor fire management. The protection of old growth and semi-natural woods and forests is a high priority of environmental and community groups throughout the region, along with the conservation of representative forest types. International, EU and national legislation supports forest conservation, but this is often hard to implement where private and public commercial interests are at stake. Examples of community involvement in the management and protection of forest ecosystems were illustrated by the potential of NTFP use in the boreal zone; fire prevention in Portugal, and the role of community involvement in saving remnants of Europe's woodland heritage.

Notes

[1] WWF & IUCN (1994): *Centres of Plant Diversity. A Guide and Strategy for their Conservation. Volume 1 Europe, Africa, South West Asia and the Middle East.* WWF & IUCN, Gland, Switzerland. Also: Kuusela, K. (1994): *Forest Resources in Europe 1950-1990. European Forest Institute Research Report 1.* Cambridge, UK: Cambridge University Press. Detailed information on natural forest types is available from ESSE, EEC, EAEC (1991): *Corine Biotopes.* Brussels: European Commission. The CORINE inventory of sites of major importance for nature conservation. This has identified over 700 coniferous and deciduous forest biotopes in Europe. For example, it identifies some five different sub-types of beech forest (*Fagus* spp.), depending on the region and forest management practices: the Central European acidophilous beech forests; the Pyreneo-Cantabrian neutrophile beech forests; the Subalpine beech woods; the Southern medio-European beech forests; and the Hellenic beech forests.

[2] Halkka, A. & Lappalainen, I.(2001): *Insight into Europe's Forest Protection.* WWF International, Gland, Switzerland.

[3] Stanners, D. & Bourdeau, P. (Eds) (1995): *Europe's Environment. The Dobris Assessment.* European Environment Agency.

[4] Thanks to Sarah Lloyd for providing information on NFTP use in the European boreal zone.

[5] Salo, K. (1995): "Non-timber forest products and their utilization", *Multiple-use forestry in the Nordic Countries,* Marjatta Hytonen (ed.) METLA, The Finnish Forest Research Institute.

[6] Forest cover is rising in nearly all countries, with the largest increases in Spain, France, Portugal, Greece and Italy, with smaller increases in Ireland and the United Kingdom.

[7] Naturalness is the degree to which a forest corresponds to the original forest in terms of species composition and ecological processes.

[8] Dudley, N. (1992): *Forests in Trouble: A Review of the Status of Temperate Forests Worldwide.* WWF International, Gland, Switzerland.

[9] Stanners, D. & Bourdeau, P. (Eds) (1995): op.cit.

[10] Sekula, J. (undated): *Circumpolar Boreal Forests and Climate Change: Impacts and Managerial Responses.* A Discussion Paper prepared jointly by the IUCN Temperate and Boreal Forest Programme and the IUCN Global Initiative on Climate Change. Temperate and Boreal Forest Programme, IUCN – The World Conservation Union, Canada Office.

[11] UN-ECE/FAO (2000): *Forest Resources of Europe, CIS, North America, Australia, Japan and New Zealand.* Geneva Timber and Forest Study Papers, No.17. Main Report. United Nations: New York and Geneva.

[12] Halkka, A. & Lappalainen, I.(2001): op.cit.

[13] Adapted from Maria João Pereira (2000): "Fire Watchers in Portugal – Sapadores Florestais". In ILO (2000): *Public Participation in Forestry in Europe and North America.* Working Paper 163. Report of the FAO/ECE/ILO Joint Committee Team of Specialists on Participation in Forestry. Geneva: International Labour Office.

[14] UN-ECE/FAO (2000): op.cit.

[15] A joint work programme with the pan-European process on the Protection of Forests (see Part IV) had four objectives during 1997 -2000:
- Conservation and appropriate enhancement of biological diversity in sustainable forest management.
- Adequate conservation of all types of forest in Europe.
- Recognition of the role of forest ecosystems in enhancing landscape diversity.
- Clarification of impacts of activities from other sectors on forest biological diversity.

[16] Gray, D, (1999): "Protected Area Networks". In Grant, K. (Ed)(1999): *Europe's Forests: A Campaign Guide.* Amsterdam: A SEED Europe.

[17] UN-ECE/FAO (2000): op.cit.

[18] ibid.

[19] IUCN/WCPA (1998): *Parks for Life. Proceedings of the European Regional Working Session on Protecting Europe's Natural Heritage.* November 9-13 1997, Island of Rügen, Germany. Gland, Switzerland: The World Conservation Union.

[20] For examples from across Western, Central and Eastern Europe, see Grant, K.(Ed) (1999): *Europe's Forests. A Campaign Guide.* Amsterdam: A SEED Europe.

[21] Broadhurst, R. (1997): "People, Trees and Woods across Europe". In ILO (1997): op.cit. Also, Henwood, K. & Pidgeon, N.

(1998): *The Place of Forestry in Modern Welsh Culture and Life.* Report to the Forestry Commission. Bangor: University of Wales, School of Psychology.

22 The Woodland Trust (2000): *A Closer Look. The Woodland Trust Annual Review 1999.* Grantham, Lincolnshire, UK. Plus, Woodland Trust's various information leaflets.

23 Reiland, R.(1999): "Roadbuilding and Forest Destruction". Luxembourg Green Forest. In Grant, K. (1999): op.cit.

24 See Milieudefensie: "Stop the logging of the old-growth forest. Tackling the paper companies". In Grant, K. (1999): op.cit.

Part IV

The Institutional Context of Community Involvement in Forest Management in Europe

INSTITUTIONS

This chapter provides some comparative European-level data on important social institutions[1] which shape patterns of community involvement in forestry: tenure arrangements; policy frameworks; forest governance structures, and economic incentives. It goes on to briefly examine particular national policies and contexts. The

implications of these institutions and contexts for CIFM will be developed more fully in the light of the case studies and analysis in Part VI.

FOREST OWNERSHIP AND TENURE PATTERNS

Ownership and tenure arrangements are institutions which shape access to, and use of

Table 6	OWNERSHIP OF FOREST AND OTHER WOODLAND AREA IN EUROPE[2]	
Country	Public Ownership	Private Ownership
Austria	18%	82%
Belgium	43%	57%
Denmark	28%	72%
Finland	30%	70%
France	25%	75%
Germany	54%	46%
Greece	82%	18%
Ireland	66%	34%
Italy	34%	66%
Luxembourg	47%	53%
Netherlands	51%	49%
Norway	25%	75%
Portugal	8%	92%
Spain	22%	78%
Sweden	20%	80%
Switzerland	69%	31%
United Kingdom	43%	57%

resources. Data on forest ownership in Europe is typically divided into two broad categories: *private* forest ownership and *public* forest ownership.[3] However, these two categories contain many important sub categories, such as *common property*[4] forests, better understood as group-owned private forests; and *commune*[5] (or municipal) forests, which constitute significant areas of public ownership. Both private and public forest ownership may be overlaid with public access and/or *usufruct*[6] rights. These are often important aspects of community forestry systems.

PRIVATE FOREST OWNERSHIP[7]

About 66% of the forest in western Europe is privately owned (this rises to about 70% of forest within the EU 15). There are large regional variations in the proportion of forest privately owned. For example, the share of private ownership exceeds 70% in some countries, such as Portugal, France, Sweden, and Switzerland; whereas in other countries such as Germany and Greece, most of the forests are under public ownership, see Table 6. Private ownership ranges from farmers with a few

hectares of forest to large private companies with thousands of hectares, as in Sweden and Finland. The average size of private holdings in Europe is 10.6 hectares.[8] Most of the forest available for wood supply under private ownership is from small individually owned forests, see Appendix 1. Forest ownership patterns are changing in many countries in central and eastern Europe whose economies are in transition to forms of market economy and who have undertaken processes of forest privatisation and restitution. Shifting patterns of ownership in central and eastern Europe have created more than 1 million new forest owners since 1990,[9] see *Germany* below. Women have always owned forests in Europe and played an active role in their management, although their role is rarely acknowledged and significantly under-researched. In Scandinavia, women own almost one third of the private forest estate.[10] In the Haute-Savoie region of France, female forest owners accounted for about 32% of the property privately owned in 1997, with an average property size of about 9 hectares.[11]

Box 10   TRADITIONAL FAMILY FOREST FARMS IN SCANDINAVIA[12]

In Sweden, private individuals (families) are the dominant category of forest owners, accounting for 80% of forest land in the Southern part of the country. The private owners clearly benefit from the financial values of their ownership, but they also value the prospect that their children will inherit the land; and that forest is capital which can be cashed in the event of family needs, such as education, farm renovation or expansion. Until World War II most forest owners lived on their property and were engaged in mixed farming systems, including forestry. However, this type of management has dropped from more than 9 million hectares to less than 4 million hectares since then. Today, many farmers do not live on their forest land, but rather live near by it or even in distant towns and cities. Forestry work on their properties is usually performed by employees of the forest owner associations or by outside contractors. They also provide the forest companies with wood.

In Finland, family farms account for over 60% of the forest estate, and over 70% of forest production. Small farm forests are considered the '13th month's salary', providing useful capital for special expenditures and unforeseen expenses. Most forests are managed for their multi-functional values. Fragmentation of forests through inheritance is an increasing problem. The average size is 27 hectares, while some estimate that 150 hectares is a minimum required for livelihood based on the forest. Over 40% of forest-farm owners have employment off the farm, and many live in urban areas.

About 30% of forest land in the EU 15 is publicly owned. There are large regional variations in the percentage of public forest lands. For example, 82% of forests in Greece are under public ownership, while Portugal has only 7%, see Table 6. The average size of holding is about 1,200 hectares. Several European countries differentiate forest land belonging to public bodies (other than State) such as cities, municipalities, communes and so on. Ownership by these institutions can be of considerable importance. For example, public forest available for wood supply under this type of ownership is high in Belgium, France, Italy, Luxembourg, Portugal, Spain, Sweden and Switzerland, see Appendix 2. In principle, decentralised administration of forests provides great potential for responding to and involving communities in forest management, and providing local benefits. The case study on *French Forest Communes* illustrates some of the constraints and opportunities of this system. Since the beginning of the 1990s, several European countries have been implementing participatory procedures for public forest land, with variable success, see the case study on *Encouraging Involvement in Public Forest Management* with data from Finland, Switzerland and Denmark.

The high proportion of forest in private ownership in western Europe provides a special context for CIFM. On the one hand it acts as a constraint to CIFM. For example, in Scotland, an unequal distribution of private land ownership results in a small number of private estates owning most of the land. In Inverness-shire 179 private proprietors own almost 74% of the county.[14] Until very recently, there have been few incentives for people to plant trees because, as tenants, they would not own them. In other parts of Europe, privatisation has resulted in loss of public access to forests, and has conflicted with indigenous peoples' usufruct rights. In some countries such as Belgium, France, Ireland and the United Kingdom, public access restrictions protect landowners' rights to manage and use their land for their own goals, clearly limiting public opportunities to enjoy forest land. In some countries, such as France, communities and user associations are working to make access to private lands easier. In the UK the government provides grants to private forest owners to encourage greater public access.[15]

On the other hand, privatisation has provided opportunities for new patterns of collaboration and access in forest management. For instance, in France, the pattern of forest ownership has been significantly influenced by the egalitarian ideology of the French Revolution. The principle of equal inheritance of heirs was enshrined in the Code Napoléon of 1804 which still forms the basis of French civil law. This has led to a fragmentation but wide distribution of forest holdings.[16] Other European countries have policies which support public access on all forest land, such as the Swedish Right of Public Access, which allows recreation and gathering of forest products for personal use in private forests. The evolution of small forest owner associations throughout Europe, based on private tenure, have generated new communities of interest and patterns of collaboration. These can be seen as an alternative types of CIFM.

INTERNATIONAL POLICY FRAMEWORKS SUPPORTING CIFM IN EUROPE

A number of international, pan-European and national policies and treaties, are beginning to support sustainable forest management, and to provide a more enabling context for CIFM. At the international level, important policy initiatives include Agenda 21; the Convention on Biological Diversity; The UNCED Forest Principles; The Forest Stewardship Council's Principles and Criteria, and the European Aarhus Convention on Access to Information, Public Participation in Decision Making and Access to Justice in Environmental Matters. These are described briefly in Appendix 3.

THE PAN EUROPEAN MINISTERIAL CONFERENCES ON THE PROTECTION OF FORESTS IN EUROPE

The Ministerial Conferences on the Protection of Forests in Europe are major co-operative policy initiatives amongst European countries (38 countries including Member States of the European Union). They represent a common political commitment to the sustainable management and conservation of forest resources, as identified in Agenda 21 and in the non-legally binding Forest Principles adopted at UNCED. The various resolutions of these Ministerial Conferences are outlined in Appendix 4.

The first Ministerial Conference on the Protection of Forests in Europe (Strasbourg 1990) passed a resolution concerned with adapting the management of mountain forests to new environmental conditions. This recognised the political, environmental, socio-economic, cultural and scientific importance of mountain forests in Europe, and the role of communities in securing their sustainable futures. A subsequent White Book on Mountain Forests in Europe has been produced as part of an action plan for the implementation of the resolution.

The second conference (Helsinki 1993) agreed that: "sustainable management means the stewardship and use of forests and forest lands in a way, and at a rate, that maintains their biodiversity, productivity, regeneration capacity, vitality and their potential to fulfil, now and in the future, relevant ecological, economic and social functions, at local, national, and global levels, and does not cause damage to other ecosystems".[17] Considerable effort has gone into defining criteria and indicators of sustainable forest management in Europe through the Pan European Ministerial processes, which is beginning to provide more opportunities for CIFM.

The third and most recent Ministerial Conference on the Protection of Forests in Europe in Lisbon, Portugal in 1998, represented the growing awareness of the social and economic roles forest play in western society.[18] The subject of public participation was chosen for further exploration and research in preparation for the next Ministerial Conference in Vienna in 2003.[19] Public participation has been recognised as a key component of sustainable forest management since UNCED in 1992, and in Europe since the Helsinki conference. This development reflects profound changes in the functioning of modern democratic societies and a transformation of society's interests

towards forests and forest resources.[20] However, the participation issue raises numerous questions for Western Europe. For example, given the high number of private forest owners in Europe, can a new balance be achieved between their individual rights on the one hand, and wider social accountability on the other? And, what are the long term cost and benefits of public participation for sustainable forest management?

THE EUROPEAN UNION FOREST STRATEGY

The Treaties on European Union have no provision for a common forest policy, although the EU has agreed a forest *strategy*, which was adopted by the Council and European Parliament in 1998. The strategy is broadly based on the principles and commitments adopted by UNCED, and the pan-European Conferences on the Protection of Forests. It recognises the diversity of Europe's forests, their multi-functional role and the need for ecological, economic and social sustainability. The EU forest strategy is to be implemented through national and sub national forest programmes. However, the management, conservation and sustainable development of forests cross cut with several EC legislative initiatives, and existing policies like the Common Agricultural Policy (CAP) and rural development, environment, trade, internal market, research, industry, development co-operation and energy policies. For example, forests contribute to rural incomes and employment, and therefore constitute an important component of the EU's integrated Rural Development Policy.[21] They are important to Europe's natural environment, and their conservation and protection are the subject of specific environmental issues such as the EU Biodiversity Strategy, Natura 2000 and the implementation of the Climate Change Convention. These may include, but are not limited to projects which support CIFM. EC legislative initiatives concerned with forest conservation and sustainable development are outlined in Appendix 5.

FOREST GOVERNANCE STRUCTURES

Patterns of forest governance[22] vary widely throughout Europe, reflecting different historical contexts, administrative traditions, and perspectives on people-forest relations. In this profile, governance includes, but is not restricted to forms of public forest administration – centralised and decentralised. It also includes traditional, local governance systems; the regulations of private forest owner associations; trusts and consortia, or types of co-management systems. As we shall see in the case studies, these systems are subject to national laws and influenced by international policies described above, but provide different constraints to and opportunities for wider public involvement in forest management. They also evolve in response to changing social and economic contexts and demands.

At a broad level, European democratic institutions and processes offer a wide range of opportunities – of relative influence – for citizens to participate in political activities and influence government decision making affecting forest resources. Democratic and administrative procedures vary throughout Europe, and include elections; referenda; consultations; appeals against decisions; environmental impact assessments; public participation procedures; citizen's juries and so on.

PUBLIC FOREST ADMINISTRATION

All European countries have Forest Codes or Laws, some of which were established in the late 19th century or early 20th century, with subsequent laws and decrees which lay down the rules governing the administration and supervision of forests. It is beyond the scope of this report to examine these in detail, but some of the new national policies supporting multiple use forestry and participation in public forest management are outlined below.

All European countries have some kind of national forest authority, such as the Forestry Commission in the United Kingdom; the *Office National des Forêts* (ONF) in France, or the *Staatsbosbeheer* (SBB) of the Netherlands. Most national forest organisations are the responsibility of Ministries of Agriculture or the Environment, but there are some notable exceptions. For example, in Sweden, forests are the responsibility of the Ministry of Industry (Näringsdepartementet), a large ministry which integrates industry

with other issues such as energy, forest resources, infrastructure (roads, railways, shipping, telecom, etc.) and labour relations. In Germany, federal forest lands are under the Ministry of Finance, implying a more commercial view of the role of forests in society.

Forest administrations, and their various subdivisions, are responsible for forest planning, monitoring, policing, providing technical advice and distributing of subsidies and grants, and so on. Administrations are usually regionalised, either according to regional political boundaries (Germany), or according to ecological criteria (Netherlands). The relationship between the Regions and the ministries is not always easy. For example, in Italy, the regions, which have been responsible for agriculture and forests since the 1970s, have held two referenda which voted to abolish the Ministry.[23] The case study of the *French Forest Communes* in Part V explores the implications of a contested balance of power between central State authorities and communes for CIFM.

As discussed earlier, many public forest authorities are undergoing profound changes in their methods of forest planning and management, in attempts to make them more responsive and accountable to the wider public, and are experimenting with public participatory processes, and co-management systems. These clearly provide opportunities for new patterns of involvement in resource use.

CO-GOVERNANCE

There are a wide variety of co-governance or co-management institutions throughout the region. These are basically partnership agreements, formal and informal, between different stakeholders for the management of natural resources. Formal or legal entities include trusts and consortia.[24] These provide special opportunities and constraints to CIFM, see Box 11.

LOCAL GOVERNANCE

As discussed earlier, many traditional and communal forest governance systems were privatised or incorporated into state systems during the

Box 11   **FOREST CONSORTIA IN ITALY**[25]

Forest consortia are *partnerships* between public and private owners for sustainable forest management. In Italy they are legal entities with a voluntary membership of local stakeholders represented by groups and elected officials.

Many consortia have ancient roots in the Roman rights of communities to be users and beneficiaries of land resources. They were revised by 19th century legislation, and were nominally planned for all non-state forests areas in Italy in 1923. The first modern forest consortium was created in the *High Valley of Susa* in the province of Turin, Piedmont in 1953. The municipalities of the valley established a consortium which today is open and reinforced by voluntary membership of private forest owners. It is concerned with the planning and sustainable management of silvo-pastoral resources owned by the members, and puts considerable emphasis on protection against natural hazards. After 47 years of field work the Consortium is considered a very successful mechanism for the co-management of resources as well as for resolving land use conflicts. It builds on common cultural values, trust and consensus. It provides a means of exchanging skills, knowledge, experiences and resources among a wide range of members. And it provides a way of sharing outputs, such as revenues, employment, and protection against natural hazards. It is considered a particularly valuable tool for co-funding sustainability in environmentally and socially critical areas like mountain regions, where forests play many roles of public interest.

However, Consortia often reflect some of the tensions between central and local control of resources. For example, in the case of the Susa consortium, the Region funds most of the annual budget, while the consortium covers only one third of its costs. The centralised regional administration is more urban or polls-driven than locally or people-oriented, and this can lead to tensions over control of natural resources for rural development. Further, despite the national legislation supporting the establishment of Forest Consortia, few have really been supported by the state, and most have been created as bottom up initiatives by local users. Supporting local incomes and employment for rural development in mountain areas may not be a political priority, compared to the benefits from more commercial forest sector enterprises elsewhere. Despite their limitations, Forest Consortia are exciting interest because of their capacity to provide effective, modern, participatory governance of forest resources.

wide-scale social transformations from the 17th century onwards. As we shall see in the case studies on the *Val di Fiemme* in Italy, and the *baldios* in Portugal, some communities have struggled to retain their traditional governance institutions despite challenges from the state. Others, such as the *Crofters* in Scotland, have adapted traditional institutions for contemporary forestry purposes to good effect.

Private forest owner associations have widely developed their own systems of democratic representation, and have often evolved into companies, accountable to their share-holders. Such patterns of governance, while economically efficient and relatively transparent, can find it challenging to accept new environmental and social values and responsibilities, generated by multi-stakeholder approaches to forest governance.

ECONOMIC INCENTIVES

Financial policies and *incentives*[26] play crucial roles in shaping land use patterns in Europe in general, and in motivating community involvement in forest management in particular. The introduction of or withdrawal of financial support has significant influence on behaviour throughout the region. As we shall see from the cases studies, most types of community involvement in forest management depend, to some degree, on incentives in one form or another. See Table 7 for examples.

FORESTRY IN THE CONTEXT OF EU AGRICULTURAL SUBSIDIES

Incentives for forestry, and community forestry, in Europe should be viewed within the broader context of agricultural subsidies which shape land use and rural development patterns, and frequently put forest land use choices at a disadvantage. The most controversial subsidy shaping rural land-use throughout Europe is the Common Agricultural Policy (CAP), which was negotiated in the early 1960s at the beginning of the European Union. It is considered the Union's longest lasting and most visible sign of integration. Its objectives include: the availability of food supplies at reasonable prices; stabilisation of markets; and a fair standard of living for farmers. Over the past thirty years, agricultural practices have intensified in response to national and EC economic incentives and prompting of farm advi-

sors. Farmers have concentrated on producing greater outputs of fewer products, using more inputs such as fertilisers, pesticides and purchased animal feeds.

In the 1992 CAP reform,[28] greater consideration was given to the role of farmers in providing environmental services, such as tree planting,[29] and the use of less intensive methods of agricultural production, such as compulsory set-aside schemes for arable farmers, organic farming and general extensification schemes. However, attempts to promote tree planting on agricultural land, as a way to reduce agricultural surpluses, have not been very successful in creating economically viable wood based industries in rural areas.[30][31] Heavy subsidization levels are required to attract private owners and communities to plant trees, because of the long rotation periods in Europe.[32] Furthermore, the timber market within Europe is largely liberalised, which means that there are few regulatory or compensatory mechanisms comparable to those in the agricultural sector. Market disturbances always have direct effects on timber prices and thus on the receipts of forest enterprises. In short, this makes forestry an uncompetitive option for many landowners.

Table 7	EXAMPLES OF ECONOMIC INCENTIVES FOR COMMUNITY INVOLVEMENT IN FOREST MANAGEMENT[27]	
Type of Incentive	**Community**	**National**
Direct incentives		
Cash	Grant for Reforestation	Research grant
Kind	Office equipment	Forest access
Indirect incentives		
Fiscal measures	Lower tax rates for community enterprises	Price support for forestry
Provision of services	Rural development	Product development
Social factors	Security of tenure	Retraining of forest agency staff

The EU does provide some financial incentives for woodland establishment for grass-roots projects. For example, the LEADER Initiative, established in 1991, aims to empower rural communities and contribute to environmental integrity at a local level. LEADER I received 1.7 million ECU over 5–6 years, funding some 900 groups across Europe. The current LEADER II has a budget of EUR 1.8 billion,[33] and supports rural development investments that are managed by local partnerships. Local communities apply for money through local authorities. See Box 12 for examples of LEADER support to CIFM.

Rural areas and rural development are an important part of the EU's regional policy instruments – the Structural Funds – which are tied to objectives for specific territories.[34] The European Commission proposes to consolidate the budget of its Structural Funds over the period 2000-2006 at an annual level of EUR 30 billion for the 15 existing member states. Some policy analysts are proposing that a proportion of this budget should be used for supporting forestry and forest-based industries in marginal rural areas throughout Europe.[35]

NATIONAL CONTEXTS

Each country has its own national forestry policies, legislation and administrations. Most countries are now committed to multiple-forest use, and some administrations are beginning to support greater participation in public forest planning and management.[36] These developments are described briefly here, and are explored in more detail in several of the Case Studies in Part V.

BELGIUM

Forest policies in both Flanders and Wallonia accept the multi-functional roles of forests. The Government of Flanders Act on Forest (1990, last amended 1999) requires local forest managers to consult the local people when drafting management plans. The objectives of the public participation process have been to develop long term forest policy planning at regional and local levels; to address specific problems such as defining sustainable forest management at a forest stand level; to create new urban/recreational forests, and to improve safety and benefits for workers. The process has been organised through a series of formal and informal meetings at regional and local levels, and has involved the most concerned stakeholders such as trade unions, youth organisations, environmental organisations, industry, local government agencies at provincial and municipal levels. Public forest managers get fairly regular training although this is still often production-oriented. In private forests, public access is forbidden on the basis of private property rights. Private owners can decide themselves to open their forests to the public. In public forests, the gathering of mushrooms, fruit, etc, is tolerated if for non-commercial use (Wallonia), or permitted by special authorisation only (Flanders).

DENMARK

Multiple forest use is the main objective of forest management in Denmark. In 1995 the Danish Government decided that all 25 state forest districts should establish one or more User Councils, at local levels, to enhance involvement and influence of users in the management of public forests. Each user council has up to 14 representatives, including municipalities, major interest groups, and members elected at public meetings. They have regular meetings, but do not have formal decision making authority. The traditional forest education has focussed on economic aspects and production systems, although during the last 10 years more attention has been given to social and environmental aspects of forest management. All private forest land is open to the public during the day, but access limited to paths and roads.

FRANCE

The multiple use of forests has an ancient basis in French forest policy. It is referred to in all policy documents concerning the management of non-protected forests. At a national level, broad participation of major stakeholders helps shape long term forest management policies. Regional plans are produced by the regional commissions on Forests and Forest Products, and involve a

Box 12   **LEADER SUPPORT TO CIFM**

Mullaghmeen Forest, located north of Castlepollard, Ireland, is an expansive beech forest of over 1,000 acres. A local group have formed a co-operative and made an agreement with Coillte enabling them to develop the forest. The group have been approved for LEADER assistance towards developing walkways, car parking, sign-posting and other facilities to make the forest user-friendly. The group is also supported through a Community Employment Scheme.

In *Austria* the agricultural markets are under pressure from overproduction, and many farmers have few future income options. One alternative source of income for the agriculture and forestry sector is bioenergy supply. LEADER funds have been used to support biomass producer communities who supply centrally co-ordinated bioenergy factories. Another goal of the producer communities is to sustainably manage native forests.

The *Abriachan Forest Trust, in the Scottish Highlands*, became the owner of Scotland's largest single community managed forest in early 1999. The buyout received national press coverage for the local community who have won much commendation for their voluntary efforts with the project. The Trust have previously received funding for their initial feasibility study and subsequent business plan which were both crucial in the Trust's successful buyout bid from the Forestry Commission. LEADER support continued to help local residents undertake woodland and business management training as well as various other promotional and cultural projects.

wide range of organised stakeholder groups, such as private owners, hunters, youth groups, unions, consumers, environmental NGOs, as well as professional agencies and ministries. ONF, as well as the commune forests and towns included in the perimeter of Regional Parks, are becoming more active and there are several pilot projects at local level. The object of the participatory process is to increase the transparency of the planning process, and the acceptability of the plan by forest owners, managers and users. Some managers are more advanced in this approach than others. Private forest owners are more reluctant to open the door to forms of co-management of their forests. In principle, private forests are off limits to the general public, although many forest owners do allow public access to their woodland. Over the past decade, user communities have made efforts to improve public access to woodland by maintaining paths and creating facilities. The National Mountain Law for France (1996) also addresses special problems of mountain communities in natural resource management. It recognises that mountains constitute unique geographic, economic and social phenomena that require specific policies for development and protection. The law provides for special compensation for people living in the country's mountain regions.[37]

FINLAND

According to the new Forest Act (1997) sustainable forest management includes economic, social and ecological aspects. In Finland, the public is allowed free access to public and private forests with the rights to pick berries and mushrooms, except in some reserve areas or border zones. Hunting requires a special permit from the owner. The new Forest Act requires public participation in forest planning and management, although the Finnish Forest and Park Service (FPS) has been voluntarily implementing participatory processes for over 5 years at all planning levels – national, regional and local. Amongst others, it has been used to help define the National Forestry Programme; regional natural resources planning; nature protection and recreation planning; everyday planning at the forest stand level, and city and community land use planning. The FPS has been responsible for the implementation of the participatory processes, with the broad objective of increasing the legitimacy of the forest service by listening to owners and citizens. The participatory methods used have varied, including workshops, telephone hot lines, and public information events. The process has been open to all people.

There are about 4000 Saami people living in an area of about 30,000 square kilometres. The FPS manages over 90% of this area which contains about 600,000 ha of productive forest land as well as protected forest. The rights of the Saami to lands, waters and traditional livelihoods have not been recognised, nor put into effect in the Saami region. It is a juridically unclear situation, which threatens to estrange the Saami from their ancestors' land. The participation of the Saami in natural resource management has increased in recent years, but the possibilities to influence decisions have only rarely been on a satisfactory level.

GERMANY

German forest policy is committed to multiple-use forest management. Almost all states in Germany have plans mapping the primary functions of forests. Public participation is not required in such mapping. However, communal and state forests place considerable emphasis on the recreational aspects of forest management which always include public concerns. Private forests account for 46% of the forest area in Germany. These forests include about 600,000 hectares of Treuhand (privatisation agency) forest areas.[38] Under the German Unification Treaty, the *Treuhand* forests in the new *Länder* (state) are to be privatised unless they revert to their former owners. The Treuhand forests are those which had been expropriated between 1945 and 1949 under occupation law. The areas expropriated before 1945 or after 1949 are to be returned to their owners as far as possible on the basis of the Property Act. Forest areas that are not returned or purchased by eligible persons are to be sold at market value. The Federal Government will follow the development of forestry enterprises within the new Länder, and support all existing possibilities

for structural improvement, such as support for voluntary land exchange for the consolidation of forest areas. The public are generally allowed access to private and public forests, although hunting is strictly regulated.

GREECE

Greece has clearly stated multiple-use policies for forests with a production function. The Greek Forestry Department supports multi-purpose forest activities. Public demand is taken into consideration as it is an important factor in planning, but there are no formal procedures to involve the public. Local communities are not directly involved in forest management planning, but their needs are generally given high priority by the Forest Service. The Forest Service administers all forests – state and private. Due to lack of funds from the state, forestry professionals do not receive retraining opportunities. Forests are generally accessible to the public, except in some reserve areas.

IRELAND

In Ireland, forest planning at national, regional and local levels has involved public participatory processes, involving the Forest Service and other stakeholders. The Strategic Plan for Forestry (1996) involved a broad based consultation procedure; and at local levels participation has involved partnerships with local groups for the provision of local amenities. The distribution of subsidies to forest owners, for afforestation and recreational developments, is also open to the participation of stakeholders outside the Forest Service, including forest owners, farmers and local communities. The participatory processes increase acceptance and awareness of forest issues, help resolve conflicts and raise the efficiency of forest planning. The Irish Forest Agency uses subsidises to bring forest owners together to consider major changes in the landscape. The general public has no legal right to enter private forests.

ITALY

Forest policy in Italy is delegated to regional administrations and this causes a large variation among regions. Some regions have no policy at all. Public participation in forest planning is good in some regions, while entirely absent from others. Local communities have traditional rights to use public forests for collection of wood and other products. There are specific regional rules for hunting, and harvesting NTFPs such as mushrooms, truffles, pine seeds, chestnuts and cork – which make significant contributions to local economies. Traditional rights are usually promoted by representatives in local and national governments owning public forests. It is hard to eliminate such uses even for conservation purposes, so they are usually maintained even in National Parks. There are many different types of forest administration in Italy, from the State and regional bodies to associations of private owners and consortia. Common forest ownership also prevails in some parts of the Alps and Apennines.

NETHERLANDS

Forest policy states that all forests – where possible – should fulfil all relevant functions: recreation, production, nature conservation, landscape and environment. Public participation in the Netherlands is not particularly well developed. Only the members of some NGOs have a say in the management of their forests. Most forests are accessible to the public, as private owners receive government subsidies for forestry operations in return for providing access.

NORWAY

Forest policy is strong on social and cultural use. However, although Norway has ratified different international conventions concerned with the environment, ecological aspects of forest management are not well integrated into forest policy. Most forests are owned by private persons. All forest areas are open to the public. Public participatory processes have not been systematically developed, and are not particularly well implemented except in some areas, like big towns. There are some problems between the Saami people and small forest owners in Norway, although it appears that indigenous peoples' rights have been taken more seriously in recent years. An independent initiative for sustainable forestry – *The Living Forests Project* – was launched in

1995,[39] and involves the participation of many stakeholders including forest owners, forest industry, trade unions, consumers, recreational, women and environmental organisations. A main objective of the project is to draw up accepted and realistic criteria for sustainable forestry, and develop tools for documenting and monitoring the environmental condition of forests.

PORTUGAL

The new National Forest Act (1996) requires participatory planning at regional levels. Participatory processes have been used to undertake long term regional forest management planning, and to tackle specific projects and issues, such as the prevention of forest fires. The National forest service is responsible for the implementation of participatory processes, which has included workshops of forest owners, *baldios* (local community-owned forests), forest industries, hunters as well as the forest services.

SPAIN

Forest policy is Spain is committed to multiple forest use, although there is considerable room for improvement in implementation. Spain's National Forest Strategy (2000) was based on a public participation process lasting several years, initiated by the Ministry of the Environment. Participation is also encouraged in the National and Regional Forest Advice Councils, National and Regional Forest Product councils and Forest Defence Associations (fire protection). Public participation in the national Forest Strategy has helped establish social objectives including rural development, employment, recreation, population stabilisation in poorer rural areas. Economic objectives include production of goods, generation of income, timber for the industrial sector. Ecological objectives include care for water resources, soil erosion, atmosphere, biodiversity and landscape values. Some 100 representatives were involved in the process including stakeholders from local governments and partners, environmentalists and trade unions. The process worked through joint multi-stakeholder meetings and working groups in specific themes. Some analyses suggest that Spanish forestry participatory planning has been frustrated by technocratic approaches in some cases.[40]

SWEDEN

Multiple use forestry is not clearly defined in Swedish forest policy. Swedish forest policies have two main goals: production and care for the environment. No general systems or regulations exist that prescribe public participation in forestry matters, although the Forestry Law does contain instructions that local forest authorities should consult local community officials regarding planned final fellings in forests with specific local interests. Public forest owners do generally consult users of forests with recreational purposes before forestry operations are conducted. Consultation and consideration of Saami reindeer herding activities is a requirement in the Forestry Law for large and medium-sized forest owners in the core areas of reindeer herding. In the winter herding areas, consultation is only a recommendation for large forest owners. A large proportion of the forest land within the reindeer herding area is owned by small private owners, and consultations are scarce in these areas. The majority of the professional workforce gets regular training, which in the last few years has emphasised environmental issues. The public has free access to forests, except in some nature reserve and military areas. Berries and mushrooms are picked for local consumption and sale.

SWITZERLAND

Multiple use is ordained for all forests and generally well implemented. The New Forest Law (1992) obliges cantons to organise public participation while developing long term forest management planning at a regional level, above forest ownership. The cantonal forest services are responsible for the implementation of participatory processes, such as defining the size of the region, participatory methods, etc. The objectives have been to enhance democratic legitimacy; to obtain legal protection and increase the efficiency and efficacy of planning. The main participants have

been forest owners (public and private); social interest groups for regional forests and wider population. The forest law gives no formal requirement about how to implement participation. The cantons are free to choose their own methods, such as working groups, public meetings, letter inquiries, exhibits. The cantons are formally required to answer the public's suggestions and comments. Some local Swiss regulations governing community use of communal forest and pasture date back to the 15th century and are still valid today.[41]

UNITED KINGDOM

The UK has a clear policy statement on multiple use forestry. Open access to state forests exists and multiple use is encouraged through the Woodland Grant Scheme (WGS) and other Forestry Commission guidelines for private forests – although some environmentalists argue that biodiversity is not given sufficient consideration. All WGS applications and forest design plans are subject to public consultation and public opinion is taken into account prior to approval. Grant applications are published in local newspapers, and all people have the right to comment. There is also a public register of WGS applications. This type of involvement serves to enhance the legitimacy of public funding and help minimise local conflicts. The Forest Service handles the grant applications and is a 'broker' between parties. They do not necessarily initiate processes, or build partnerships between stakeholders. In state owned woods, permits are available for a range of activities such as fuelwood gathering, motor rallies, etc. There are growing numbers of self-mobilised community forestry initiatives in the UK, in which local people and NGOs either purchase or take on the management of forests themselves. A 1995 Rural White Paper stated that the government wished to enhance the contribution forestry could make to sustainable communities. The public has access to only a small proportion of private woodland, where there is a public right of way, or where access is promoted through a grant scheme.

NOTES

[1] See glossary for definition.

[2] UN-ECE/FAO (2000): *Forest Resources of Europe, CIS, North America, Australia, Japan and New Zealand Geneva Timber and Forest Study Papers, No.17.* New York and Geneva: United Nations.

[3] See glossary for definition.

[4] See glossary for definition.

[5] See glossary for definition.

[6] See glossary for definition.

[7] UN-ECE/FAO (2000): op.cit.

[8] Some countries that provided statistics to the UN-ECE/FAO (2000) report on private forest ownership did not include holdings under 3 hectares.

[9] FAO/ECE/ILO (1997): *People, Forests and Sustainability. Social Elements of Sustainable Forest Management in Europe.* Geneva: International Labour Office.

[10] Ekberg, K. (1997): "How to Increase the Participation of Women in Forestry – Ideas and Ongoing Work". In FAO/ECE/ILO (1997): op.cit.

[11] Faugère, I. (1999): *The Role of Women on Forest Properties in Haute-Savoie: Initial Researches.* Geneva Timber and Forest Discussion Papers No.13. New York and Geneva:United Nations.

[12] Thanks to Harri Karjalainen from WWF International for this data.

[13] See glossary for definition.

[14] Scottish land ownership patterns are rooted in feudal tenure, which has its origins in the political economy of 11th century, see Wightman, A. (1996): *Who Owns Scotland.* Edinburgh: Canongate Books. The Scottish landscape is further shaped by the effects of the forced eviction of thousands of families during the Highland Clearances – when an estimated half a million people were driven from their traditional lands, see Prebble, J. (1963): *The Highland Clearances.* London: Penguin.

[15] UN-ECE/FAO (2000): op.cit.

[16] Bedel, J. & Brown,D. (1998): "France". In Shepherd, G; Brown,D; Richards, M; & Schreckenberg, K. (Eds) (1998): *The EU Tropical Forestry Sourcebook.* London: The Overseas Development Institute.

[17] Ministry of Agriculture and Forestry of Finland (1993): *Ministerial Conference on the Protection of Forests in Europe, 16-17 June in Helsinki.* Helsinki: Finland, p. 4.

[18] FAO/ECE/ILO (1997): op.cit.

[19] ILO(2000): *Public Participation in Forestry in Europe and North America. Working Paper 163.* Geneva: International Labour Office.

[20] ILO (2000): op.cit.

[21] MCPFE (2000): *The Role of Forests and Forestry in Rural Development – Implications for Forest Policy. International Seminar, 5-7 July, Vienna, Austria.* Liaison Unit, Vienna: Ministerial Conference on the Potection of Forests in Europe.

[22] See glossary for definition.

[23] Navone, P; & Shepherd, G. (1998): "Italy". In Shepherd, G; et al. (Eds) (1998): op.cit.

[24] See glossary for definitions.

[25] Thanks to Pier Carlo Zingari for this example.

[26] See glossary for definitions.

[27] This table is adapted from McNeely, J.A. (1988): *Economics and Biological Diversity: Developing and Using Economic Incentives to Conserve Biological Resources.* Gland, Switzerland: The World Conservation Union.

[28] The CAP was again reformed in early 1999. The aim of the new reforms is to continue changes introduced in 1992 to divert aid away from price subsidies towards direct support for farmers. Rather than buying up farm products to keep prices artificially

high, the changes aim to extend the policy of compensating farmers directly for losses. The reform is also aimed at bringing the EU budget under control to set the scene for new members to enter the union in the coming years. It is also intended to equip the EU with a position it can defend in world trade liberalisation talks. The subsidies drive prices higher than the world level, and products have to be subsidized further to compete on global markets. The EU is under pressure from the US and Australia to reduce market distorting farm aid. Agricultural support is the biggest element in the EU 86 billion Euro annual budget. New reforms call for spending to be kept at its present level of 40.4 billion Euros. A contested issue within the European Union is that some 80% of the CAP budget goes to 20% of the farmers (40 billion ECU to 1.4 million farmers). Environmental NGOs, such as WWF, have been lobbying for more environmental safeguards within the CAP.

[29] See Appendix 5, Regulation 2080/92.

[30] European Environment Agency (1999): *Environment in the European Union at the Turn of the Century.* Copenhagen: European Environment Agency.

[31] Over 500,000 hectares of land has been afforested under this scheme – more than half of this total has been in Spain.

[32] Stanners, D. & Bourdeau, P. (1995): *Europe's Environment. The Dobris Assessment.* Copenhagen: European Environment Agency.

[33] 1996 prices. The former ECU has been converted into the EURO.

[34] The EU regional policy instrument – the Structural Funds – pertain to several rural areas: Objective 1 regions are those lagging behind economically, with a GDP of less than 75% of the EU average. Objective 5a regions receive support for downstream processing of agricultural produce as well as investment aid and compensatory allowances in less favoured areas. Objective 5b applies to rural areas with a low level of socio-economic development, high dependency on agricultural employment, low agricultural incomes and population problems such as low density or declining population. Objective 6 applies to regions north of the 62nd parallel with a very low population density (less than 8 inhabitants per square km).

[35] Glueck,P. (1997): "Sustainable Forestry in the Context of Rural Development". In FAO/ECE/ILO (1997): op.cit.

[36] Unless otherwise stated, the information in this section is drawn from three main sources:
●WWF (1998): *European Forest Scorecards 1998 Report.* Gland: WWF International;
●UN-ECE/FAO (2000): *Forest Resources of Europe, CIS, North America, Australia, Japan and New Zealand. Geneva Timber and Forest Study Paper No.17.* New York & Geneva: United Nations;
●Shepherd, G; Brown,D; Richards, M; & Schreckenberg, K. (Eds) (1998): *The EU Tropical Forestry Sourcebook.* London: The Overseas Development Institute.

[37] Messerli, B. (1999) in Lynch, O.J. & Maggio, G.F. (2000): *Mountain Laws and Peoples: Moving Towards Sustainable Development and Recognition of Community-Based Property Rights.* Harrisonburg: The Mountain Institute; Washington DC: Centre for International Environmental Law; Gland, Switzerland: Mountain Forum.

[38] Federal Ministry of Food, Agriculture and Forestry (BMELF)(undated): *Forest Report by the Federal Government.* Bonn: BMELF.

[39] *The Living Forests Project Facts*, P.O. Box 1438, Oslo, Norway.

[40] Garcia Pérez, J.D. & Groome, H. (undated): *Spanish Forestry Planning Dilemmas: technocracy and participation.* Mimeo: Department of Environmental Management, University of Central Lancashire, Preston, UK.

[41] Messerli, B. (1999) in Lynch, O.J. & Maggio, G.F. (2000): op.cit.

Part V

CASE STUDIES OF COMMUNITY INVOLVEMENT IN FOREST MANAGMENT

INTRODUCTION

This section presents 12 case studies of community involvement in forest management from around Western Europe. The aim of this section is to illustrate and learn from the diversity of ways in which rural and urban peoples are involved in forest management within the region. See Figure 5 for location of case studies

THE FORESTS OF THE VAL DI FIEMME IN NORTHERN ITALY: TRADITIONAL INSTITUTIONS SUPPORT THRIVING COMMUNITIES[1]

HISTORY OF COMMON FOREST MANAGEMENT.

There are several long established traditions of communal forestry in certain valleys of the eastern Alps of Italy, such as *Val di Fiemme, Comelico, Conca Ampezzana*, in which communities jointly own and collectively manage the land. Evidence of common forest management can be found in numerous statutes dating from the Middle Ages and relate to both Alpine and Apennine village communities. Forest and pasture management was at the heart of the community's social and economic life in the mountain areas. Forests and pastures were characterised by their inalienability, indivisability and collective management. Historical research indicates that the roots of these institutions can be traced back to prehistoric settlements existing well before the Roman conquest.

Thanks to the wealth generated by forestry in the Middle Ages, some of these communities rose to the status of 'Rural Republics' which were largely independent from feudal lords and the cities. Wood was distributed according to a member's needs for building a house (once in a lifetime) and for maintenance work and heating (once a year). The financial returns from the sale of surplus timber was used to support the organisation of the village community, especially to assist the poor and to provide health care, education, road construction and maintenance, water supply and emergency funds. Since the 1950s the benefits from forestry have been used to mitigate the shocks of rapidly changing society, particularly in areas with high rates of emigration and the breakdown of family units which resulted in children, women and the elderly being left behind in villages. The revenue from communal forests has been used to develop new activities providing an incentive for people to remain in rural areas.

As discussed in Part II, the establishment of modern states in the seventeenth and eighteenth centuries throughout Europe, favoured the centralisation of authority and the creation of public and private property, challenging the common forest properties of village communities. In 1927 the Italian Government passed an act aimed at enclosures and liquidating collective property and transferring it to the new administrative communes. Special attention was given to the gradual dissolution and abolition of the *usi civici* (civic customs)

Figure 5

LOCATION OF CASE STUDIES

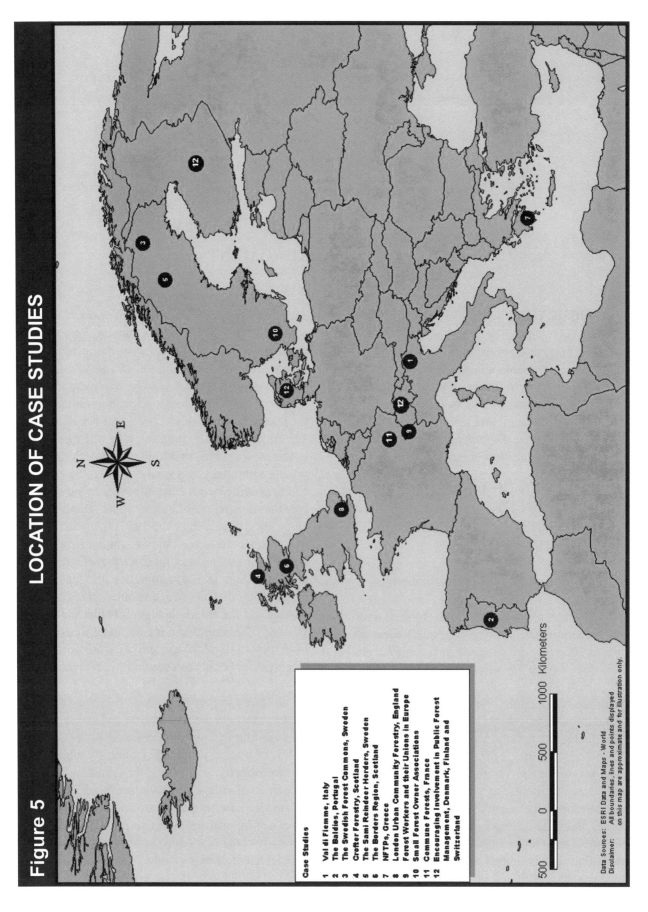

Case Studies

1 Val di Fiemme, Italy
2 The Baldios, Portugal
3 The Swedish Forest Commons, Sweden
4 Crofter Forestry, Scotland
5 The Sami Reindeer Herders, Sweden
6 The Borders Region, Scotland
7 NFTPs, Greece
8 London Urban Community Forestry, England
9 Forest Workers and their Unions in Europe
10 Small Forest Owner Associations
11 Commune Forests, France
12 Encouraging Involvement in Public Forest
   Management, Denmark, Finland and
   Switzerland

Data Sources: ESRI Data and Maps - World
Disclaimer: All boundaries, lines and points displayed
on this map are approximate and for illustration only.

which were present in almost all regions of Italy. However, communal forestry was able to survive in more remote areas where it was more deeply rooted, or where resources were perceived to be less economically important to the state, although not to the local population. Some communities struggled hard to stop the dissolution of ancient community rights, and the transfer of their lands to the municipalities, discussed shortly. In general, modern state structures and economic development have progressively reduced the role and expanse of the village communities, particularly in the past 200 years. Today they have practically disappeared and cover only some 200,000 hectares or 5% of the country's alpine forests. However, in areas where a solid economic base is supported by deep rooted ethical and cultural values, communal forests have survived and thrived.

COMMUNITY INSTITUTIONS

Historically, written rules regarding rights and duties were laid down to regulate the social and economic life of village community members. The *regola* (literally 'rules') consist of institutions and define and assign duties and rights, relating to both political and economic domains. At the base there were family communities, or groups of interrelated families who collectively owned and managed the land at the hamlet level. Then came the village communities which were in turn grouped into Federations. Some of these were large, consisting of members from a group of valleys such as the *Magnifica Comunità of Cadore* or *Val di Fiemme*. Direct democracy was applied to such an extent that the leaders were not elected but rotated each year from among the heads of the village families

to avoid oligarchies. Some communities were restricted to original members, so that membership was acquired as a birthright through the male line, while others were open to new members. The forest land and pastures had strict rules regulating their use to ensure continuity of production. Collective ownership of forest lands and pastures in marginal alpine areas has been characterised by a strong spirit of mutual assistance and solidarity, which has provided an important cultural basis for the use and development of resources.

THE VAL DI FIEMME, TRENTINO

Most of the forests of the *Val di Fiemme* in north east Italy are owned and managed by a variety of community organisations, see Figure 6 and Table 8. The valley covers nearly 50,000 hectares of land, with an altitude range of 630 to 2842 metres. Forests cover 29,000 hectares (60% of the land surface), and consist mainly of spruce stands *(Picea abies)* (over 80%). Spruce finds its optimum habitat in the valley, and individual trees often obtain remarkable heights of 45 metres. In general, the Fiemme spruce stands are even aged, with a growth rate of about 5-6 cubic metres per year. After tourism, the forests provide the main source of income, employing about 800 workers in some 200 companies, out of a total population of about 20,000. One of the main reasons why the forests have continued to play an important role in the economy of the Val di Fiemme is the presence and activity of the Magnifica Comunità di Fiemme (see below) and the key role of the villages in the direct administration of the woods. The selling of timber by the road has helped maintain a number of local wood logging and small wood enterprises.

Table 8	FOREST PROPERTY OWNERSHIP, VAL DI FIEMME, TRENTINO, ITALY
Magnifica Comunità di Fiemme	40%
Village-owned woods	40%
Large private woods (single or collective)	8%
Province-owned woods	7%
Small private woods	4%

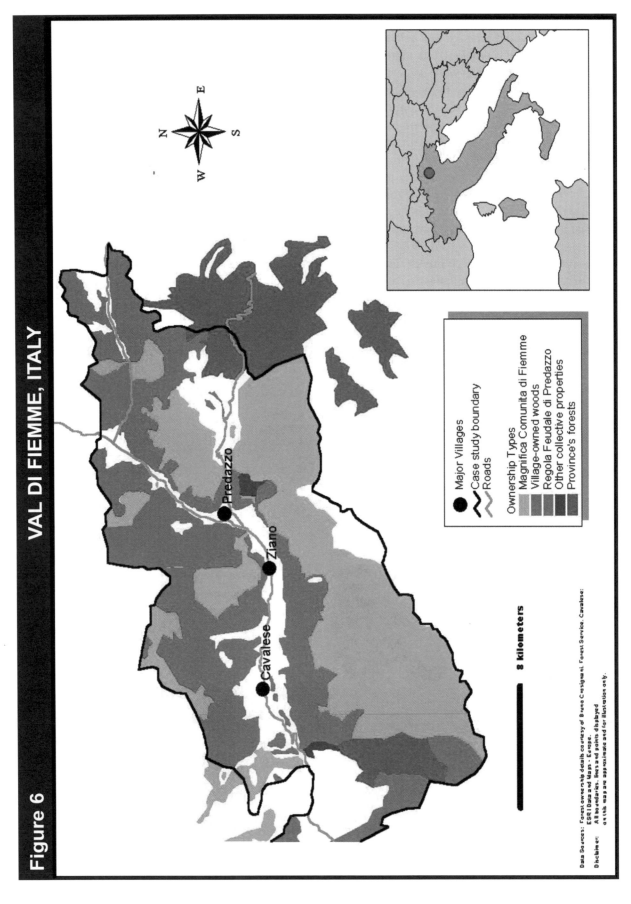

Figure 6

VAL DI FIEMME, ITALY

Predazzo

Ziano

Cavalese

Major Villages
Case study boundary
Roads
Ownership Types
Magnifica Comunita di Fiemme
Village-owned woods
Regola Feudale di Predazzo
Other collective properties
Province's forests

8 kilometers

Data Sources: Forest ownership details courtesy of Bruno Crosignani: Forest Service. Cavalese:
ESRI Data and Maps - Europe.
Disclaimer: All boundaries, lines and points displayed
on this map are approximate and for illustration only.

The general rule in other parts of Italy is to sell timber standing in the woods. According to traditional citizen's rights, the village owned (or municipal) lands and forests, such as those in Tesero and Cavalese, provide the local population with fuelwood, building timber (or financial equivalents), and pasture.

THE MAGNIFICA COMUNITÀ DI FIEMME

This is one of the most important examples of community ownership in the Italian Alps. The Magnifica Comunità is recognised as a unique institution – a remnant of an ancient public body, which was first officially recognised in 1111 by the Prince Bishop of Trento. He granted them relative autonomy, allowing the people of the valley to rule themselves according to their own customs and laws, except in important matters of justice, in exchange for 24 prince's soldiers per year. Their ancient autonomy has been endangered on many occasions, till in 1951 their rights and autonomy were codified in a new statute.

The community owns about 20,000 hectares of land, with the following resources:

Forest	11,000 ha (mainly spruce)
Alpine Pastures	5,000 ha
Pastures	2,000 ha
Unproductive land	1,000 ha

The land belongs to the *homines Vallis Flemarum* – to all the people of the Valley of Fiemme. The inhabitants *(vicini)* of eleven townships share the possession of the estate. According to a 1993 statute, a *vicino* is a person who has been living in the valley for 20 years or who is a descendant of a *vicino*. The Comunità consists of a Town Assembly, composed of representatives elected in each town; a General Assembly, or representatives of those elected in the individual towns (the number is proportional to the vicini with a minimum of three representatives for each town); the Council, or real executive organ composed of eleven people, one from each town; the Control Board, appointed by the General Assembly to control and review the administration and accounting acts, whose members are external to the General Assembly; and the President *(Scario)* and Vice President, appointed by the Council.

The forests, managed by two foresters and nine wardens, provide about 46,000 cubic metres of wood per annum which is processed at their own sawmill at Ziano, see Plate 7. The Comunità made the decision to apply for independent FSC (Forest Stewardship Council) certification in 1996, after hearing about the system at a World Forestry Congress. The Comunità di Fiemme applied to an English body (SGS in Oxford), and after two assessment visits was issued its certificate in 1997. Only minor changes were required for the Comunità's forests to fulfil the ten obligatory social, economic and ecological standards set by the FSC. Independent certification has since provided excellent business opportunities for the Comunità, and increasing demand for FSC wood from European buyers currently exceeds available supplies from its sawmills at Ziano.

The sawmill at Ziano produces industrial wood products, while other manufacturers and carpenters produce semi-finished products which supply larger industries, furniture, wood boxes, handicrafts, and musical instruments. Revenues from communal forests is still used to provide public services. The recent introduction of modern technology into the Ziano sawmill has transformed wood processing in the area and made the industry one of the most up to date in Italy. The sawmills currently process about 35,000 cubic metres of logs for planking for industrial products. The industrial upgrading of the sawmills introduced at the end of 1990, such as stripping bark of logs, and measuring and classifying logs, has enabled the company to radically improve its share of the market. The semi-worked products section is currently expanding and has brought about a significant increase in quality. The Ziano saw mill employs about 55 local people. There are about 30 felling companies, and 15 transport companies working for the Magnifica Comunità.

THE REGOLA FEUDALE OF PREDAZZO

This is another community within the same valley owning communal forest and agricultural holdings which provide an important basis for the local economy. The *Regola Feudale*

PLATE 7: The Saw Mill at Ziano.
(Photo: J-P. Jeanrenaud)

(Feudal Rule) of Predazzo is an ancient village institution recognised by the Prince-Archbishop of Trento in 1447. In the past, Predazzo was the smallest village of the Fiemme Valley with just a dozen families. Besides enjoying some rights on the large estates of the Comunità, they have concentrated their agricultural activities, including forestry, on the northern slopes of the upper valley. The land provided pasture for livestock and forests for firewood, structural timber and so on. Over time the members of the *Regola* increased, but nobody except descendants of the original families was admitted. The first statutes were drawn up in 1608 and had seventeen articles, signed by the sixty one members *(vicini)* of the *Regola*. These related to rights of exclusion, and regulations concerning the use of pastures and forest, time of harvest, transport and so on. The statutes have been modified over the years. The *Regola Feudale* of Predazzo underwent a long series of civil trials,

decisions and appeals over some forty years, until a final decision of the Superior Court in 1967 declared that the "*Regola feudale* is a true private community, the members of which have in common the ownership of the land". In early 2000 the *Regola* had some 19 families, with about 800 individual *vicini* spread all over the world. The members of the *Regola* succeed through the male line. This rule was challenged in court in the 1990s, but the traditional rule was legally upheld.

As described in its statutes, the *Regola Feudale* of Predazzo is a community having private rights in common to the ancient agro-silvi-pastoral patrimony. The aims of the community are the conservation and management of the resources under the community, involving animal husbandry and forest resources, according to national forest regulations; and to promote work opportunities for the families of the vicini. The

66

Regola is managed and administered by a General Assembly who elects 9 members of a Council, and by a *Regolano* (President), elected by the Council, who manages current affairs. Conflicts of interest between the *vicini* sometimes arise, such as recent differences between those who wish to develop the ski facilities of the region and those who wish to retain more traditional resource use. Difficult issues are brought before and managed by the General Assembly.

The present patrimony of the *Regola Feudale* covers 2698 hectares of land, with:

Forest	1, 156 ha
Alpine meadows	677 ha
Alpine ranges	362 ha
Arable land	6 ha
Non-productive land	404 ha

The forests extend from the bottom of the valley at 1000m, to the upper limit of forest growth at 2200m. The main species is spruce 81% *(Picea abies);* larch 16% *(Larix decidua),* with some *Pinus cembra, Abies alba,* birch moun-tain alder and willow. Because of the steepness of the land, all the forests fulfil an important protective function against soil erosion. The forests are managed according to a management plan approved by the Province's Forest Service, and in agreement with the local Provincial Inspector. Mangement plans are subsidised by the Province. Management follows a combination of selective and shelterwood systems to encourage uneven aged stands. The total prescribed harvest is 3,600 cubic metres per year. The *Regola* employs one forest guard and about 5 workers during the summer. Harvesting and exploitation of the cut is contracted out with priority to the *vicini.*

In the past, rangeland, meadows and arable land were important part of the basis of the rural economy of the *vicini.* Very detailed and strict rules applied to these lands. Over the last fifty years the economy of the valley has changed, and agriculture has been largely abandoned in favour of the wood working industry and tourism, including skiing. The *Regola* owns and manages about 50 km of ski track. Animal husbandry is still quite important but restricted to a few rural families.

PLATE 8: Household fuelwood provided by communal forests.
(Photo: J-P. Jeanrenaud)

The *Regola's* income amounts to some US$ 500,000 per annum, including income obtained from forest resources, rent from ski lifts and tracks, mountain huts, houses and offices. About two-thirds of the profits are distributed among the *vicini* (about US$ 100 per annum), while the other third is reinvested in improved infrastructure. While individual income is small, sharing the profits helps maintain a sense of community and interest in common resources. The woodworking industry, especially handicraft production, and trade in wood play an important role in the economy of Predazzo today. A significant proportion of the village economy now depends upon wood exploitation and utilisation. In the past, fruit crates were one of the main products, but these days furniture, windows, doors, flooring and pallets are being produced. Traditionally, the forest provided construction and firewood, and wood for agricultural tools. *Vicini* are still permitted to collect several steres of fuelwood per year.

OTHER COMMUNITIES

Revenue from timber and pastures in other communities has declined since the 1980s, while costs have increased as a result of management aimed at conservation and the provision of recreational services. These days, communal forestry has shown its ability to play new roles in mountain regions, particularly in the fields of tourism and prosperous small industries. Two important institutions, the *Regole di Cortina* and the *comunalie parmensi*, today present balance sheets in which timber sales are only one third of the total revenue. In the former case, access rights for sports and tourism -skilifts, slopes, car parks, camping grounds and other rents – are the main revenue, while in the latter case, it is medicinal herb and mushroom harvesting rights. These rights are sold as day permits and have increased from 4000 lire (US$ 2.70) in the early 1980s to more than 10,000 lire (US$ 6.70) today while the number of visitors has remained stable or even increased. Under traditional property rights, the collection of NTFPs, hunting and recreational activities were (and so some extent still are) unrestricted and free of charge. However, the growth in tourism and increased demand for environmental

services highlights the need to define and assign use rights over NTFPs and environmental benefits in communal forestry today.

LESSONS LEARNED

◆Community forestry institutions – where they have survived – have helped provide a source of economic and social wealth in many regions of north east Italy. Political autonomy, combined with economic enterprise, strong social ties and interest in community welfare, have helped create and sustain robust and thriving communities.

◆Political, economic and social changes over the last few centuries – such as the centralisation of state authority, the privatisation of land and rural emigration – have undermined community forest institutions in many areas. Regions with stronger social ties and financial resources, like those in the *Val di Fiemme*, have managed to resist the trends towards greater state control and privatisation of forests. Several communities have successfully faced legal challenges to retain their independence and autonomy before government legislation became more supportive during the 1950s.

◆Community forestry allows for economies of scale (over thousands of hectares) for sustained timber production and effective multipurpose management. These economies of scale are often not possible in individually owned private forest estates, which are generally rather small in Italy. Community forestry demonstrates a higher degree of dynamism and flexibility in the face of social and economic change, compared to other types of land tenure.

◆In order to survive, communities have been obliged to adopt market solutions. Communities have benefited by upgrading wood processing technologies, and adopting new initiatives, such as selling timber by the road, and FSC certification. The shift to commercialisation of NTFPs and environmental goods has also helped communal forestry remain viable amidst social and economic changes.

◆Communal forestry institutions allow for wider local participation in forest management decision

PLATE 9: Brush from the *baldios* is used as bedding in the stables, and then used to fertilise the fields in spring. *(Photo: R. Brouwer)*

making. Democratic assemblies of members prove effective instruments for finding locally acceptable compromises in natural resource management. However, there may be risks of elitism where collective property has a legally 'private' status. It excludes those who do not originate in the area, while at the same time maintaining the rights of families who emigrated generations earlier. Where legally recognised as 'public', and open to all residents, there are fewer concerns of this nature. There may also be problems of gender bias if rights are inherited only through the male line.

THE BALDIOS IN PORTUGAL: COMMUNITIES RECLAIM THE COMMONS[2]

THE BALDIOS

Many of Portugal's *baldios* (or commons) still provide rural people with important community forestry benefits. The *baldios* are an ancient

tradition in Portugal, dating from at least the Middle Ages. They were once mainly used as pasture and as a source of fertilizer. Today they are a source of marketable timber and resin, as well as providing traditional resources. However, the history of community involvement in the management of the *baldios* has had a chequered past and is frequently challenged by the state as well as wider social and economic change. In the late nineteenth century, the *baldios* comprised more than 4 million hectares of land, but by the advent of the new state regime *(Estado Novo)* in 1933, privatisation had reduced them to some 450,000 hectares (or about 5% of the land surface of Portugal). These remaining lands are concentrated in the north and interior of the country.

TRADITIONAL INSTITUTIONS

The *baldios* played a key role in traditional farming system, where they provided construction material, fuel, pasture for goats, sheep and cows, bedding for stable animals. They also provided a source of green and brown manure for private cultivated fields. In many cases individual households held their cattle in communal herds. The *baldios* were the material base of community organisation and sophisticated structures existed for the administration of common resources. The main components of the traditional management system were exclusion, zoning and allocation. Only people who were accepted by the community were permitted access to the commons. The *baldios* provided a form of social security for the poor, who were permitted to graze pasture cattle and to cultivate plots on a temporary basis. Village councils or *chamados* determined whether areas were open to grazing or not, which areas were reserved for regrowth, and which parts could be used. Implementation of the council's rulings was supervised by elected caretakers or *zeladores*.

CONFLICTING PERCEPTIONS

Government backed privatisation, and land usurpation greatly reduced the area of the *baldios*. They were widely perceived as being unused or being used in undesirable ways. The term *baldio* is derived from a Teutonic idiom meaning barren, waste or bald. These lands have always been a bone of contention between the rich and the peasants, between shepherds and farmers and between local groups and central authorities – including the forest service. Under the reforestation policies of the new state regime about 300,000 hectares of *baldios* were planted with trees, mainly *Pinus pinaster*, and placed under the management of the state forest service. This effectively curtailed many traditional usufruct rights, and forced many community members to emigrate and seek other occupations. The state forest service and its afforestation efforts were viewed with hatred and resisted by the local populations despite the fact that some government compensation was promised when the timber was harvested. In some cases forestry projects could only be implemented with police protection. Tree planting was effectively part of the subjection of the local population to state law and power. In 1966 the civil code was revised to abolish communal property, and for the next decade, the *baldios* ceased to

PLATE 10: A private field with communal lands on the hills.
(Photo: R. Brouwer)

exist, at least offically. In 1974 a leftist-inspired military coup ended the new state regime, and along with it state control of the *baldios*.

RESTORATION OF THE BALDIOS TO THE POPULATION

In 1976 a law was passed aimed at the restoration of the commons to the commoners. The law returned the land to the original user-communities rather than to municipal councils, partly on the advice of the Forestry Service, to avoid breaking up the forested areas. In many villages people created commoners assemblies and management councils to undertake the administration of their traditional domain. These became the formal legal democratic representatives of the village in common land affairs. The commoners' assembly was required to elect a five-member council responsible for day-to-day management of the commons. 84% of the newly formed councils opted for managing their lands in cooperation with the state. The 1976 law brought about the creation of 637 councils, which claimed their commons back from the forest service and municipal councils. However, since that time, many of these have ceased to function or have been dissolved. Today only 132 are still operational, and only about one third of the commons (141,000 ha) are actually administered by local communities.

CHANGES IN RURAL AREAS

Since the 1930s and the era of state control, many important changes had taken place in the countryside. Many of the commons had been forested by the state administration resulting in the availability of timber and resin. The 1976 legislation tried to harmonise traditional commoners' rights with forestry management of the *baldios*. There had been significant rural depopulation since the 1960s, largely as a result of the closure of common lands and the consequent decline of animal husbandry. When the commons were opened again, there were no people or animals left in many areas. Chemical fertilizer was now widely used by rural people, meaning that local people were no longer fully dependent on brush from the commons for manure. Furthermore, there had been a tendency to abandon farming altogether, partly as result of the EC's CAP, which

has favoured modernised farms at the expense of 'marginal' farmers. Between 1970 and 1990 the percentage of the working population in primary agricultural production dropped from 32% to 18% on a national level, and from 70% to 50% in the north. Together, these trends have affected the value of the commons and the way they need to be managed. Traditional forms of exploitation of the commons have become less important, while new forms related to the presence of trees have gained in importance.

RECENT LEGISLATION

Since 1976, some 17 bills have been presented to parliament, aimed directly or indirectly at the dissolution of the user communities and the transfer of their management powers to the municipal councils. However, these have been thwarted by a lack of consensus among the powerful parliamentary parties. However, in the early 1990s, a compromise was reached and a new law was published in September 1993, leaving intact the idea of local administration by the commoners. The management councils are to consist of community members only, without a state representative, except in cases where councils have chosen to cooperate with the state. A new commission, charged with the supervision of the management council's financial administration, has been created. The new law also allows the administration to be delegated to another entity (e.g. municipal council or relevant government office such as the forest service); expropriation by the state for the public's good; privatisation for the benefit of housing or industry and the extinction following unanimous decision by the commoners themselves or after three years of 'ostensive abandonment'. In agriculture it is not unusual to leave land in fallow for seven years or more, suggesting that it is too early to say whether common land has been abandoned after 3 years. There is already a widespread practice of common land allocation for housing construction. This is sometimes an acceptable practice, particularly when common land is used for the construction of housing for villagers without sufficient land. Most commoners also agree with the use of land for industrial development. However, commoners generally speak out against the abolishment of the commons altogether. The measures mentioned here pose a serious threat to the continued existence of Portugal's commons.

CIDADELHE DE AGUIAR

This village lies 20 kilometres to the north of the Vila Real district. It manages a common of some 700 hectares, and has assumed full responsibility over the area, so that the state is no longer represented on the management council. The common was forested between 1945 and 1965, a heavy blow to the local economy which depended almost totally on sheep and goat farming. Despite former hardships resulting from afforestation, the forests now offer large profits to the villagers. The community, of about 135 inhabitants, receives about US$ 8000 annually from the sales of resin alone. It earns money from occasional thinnings and can expect revenue from the first cuts within a few years. Between 1986 and 1989 the council administered an annual gross revenue of US $ 25,000. This has been invested mainly in infrastructure for the public benefit, such as improving the agricultural irrigation system, construction of footbridges and a community centre. The council has contributed to the construction of a football field, and subsidizes club membership fees for the younger players. The council assumes all the silvicultural tasks that would normally be carried out by the state, including organisation of thinnings, felling, resin collection, and tending. It acts as a modern forest entrepreneur, but has the obligation to maintain the forest cover according to legal requirements. It also pays the forest service 30% of its share in the gross timber revenues, The forest service provides advice.

AVEÇÃOZINHO

This is a village within the district of Campeà. In the 1950s the state intervened by allotting 14 ha of the commons to 20 of the 58 households in the village, and by foresting about 50 ha. The remaining 20 ha. of the common land had to be used by some 38 families. When the commons were returned to the community in 1976, the commoners established a management council for their administration. Although formally responsible for the whole commons, in reality the council only exercised authority over the 20 ha. that had not been under the control of the forest service. State management of the forested part of the commons had been of little benefit to the local people. In the early 1980s the pine planted on commons burned down before reaching maturity. After the fire the forest service replanted some parts with douglas fir *(Pseudotsuga menziesii)* and in 1990 opened fire corridors. These decisions were taken without consulting the village management council. The council of Aveçãozinho focussed its attention on the lowest part of the common which had become covered with pine, either planted by the people themselves, or regenerated naturally from the nearby state forest. The common land has gained importance with respect to fuelwood. With increased tree cover since the 1950s, the management council's main role is to organise the distribution of fuelwood to individual households. Each year in summer it marks the trees and distributes them through a lottery system among the commoners. This system copies an earlier system, which was used for the allocation of brush. Grazing on common land has become less important over time, although brush still plays an important part in the local farming system. Nearly all the farmers collect a cart load once or twice a week for their stables, which was traditionally regulated by the village council. Soiled brush and manure is spread on fields before ploughing and planting. These days the availability of chemical fertiliser and private brush lands have taken the pressure off the common land to provide nutrients. Some of the management functions of the council are maintained with difficulty. In the spring of 1993 the council refused to organise the allotment of trees that had been damaged by snowfall and storms. One of the villagers who had a tractor collected the damaged trees for personal benefit, much to the resentment of the other members of the community.

LESSONS LEARNED

◆ As in many parts of the developing world, the contribution of common land to rural economies was perceived as undesirable or insignificant by the Portuguese state for many years, and land was appropriated on a wide scale and local institutions dissolved, with disastrous implications for local livelihoods.

◆Communities were able to take advantage of the 1976 law which returned the management of the *baldios* to democratic assemblies of local people. However, local management councils required legal recognition and protection from the state. There have subsequently been a number of challenges to the continued existence of the *baldios*.

◆Despite initial hardships resulting from state appropriation of the commons, the forested *baldios* now provide communities with important benefits such as fuelwood, and revenues from marketable timber and resin.

◆Traditional management institutions – which regulated animal grazing and collection of brush – can be adapted for the management of forest resources. Communities now organise auctions; negotiate with concessionaires, traders and the forest service; invest revenues, and achieve good market prices for forest products for the benefit of the community.

SWEDISH COMMON FORESTS: ADAPTATIONS TO URBAN, INDUSTRIALISED SOCIETY[3]

INTRODUCTION

There are about 33 forest commons in Sweden today, encompassing a total forest area of about 730,000 hectares, owned by over 25,000 people who hold 'shares' in the forest resources. The Swedish forest commons date from the late 19th century, but some were modelled on ancient medieval systems. When the commons were first created, the forests had little monetary value. Today, many commons are run like big forest companies. The three largest each possess about 60,000 hectares of productive forestlands, and have a sizeable workforce and technical infrastructure, providing their shareholders with substantial benefits. How are these common forests organised, and how have they adapted to modern industrialised society?

HISTORICAL DEVELOPMENT

Different types of commons were codified in old country laws in Medieval times. In 1523 Sweden became independent and was united under King Gustav Wasa. Huge areas of land were confiscated from the nobility and from the church. Under the monarchy, all 'unclaimed' land was declared to be the property of the Crown/State. In the 17th century a process of land delimitation was initiated, followed by a process of land redistribution in the middle of the 18th century. These procedures created bigger and more productive farms and widened the basis of taxation. As a result of land delimitation, farmers were allotted their own private forests, many of which had only minor value at the time. In the early period of industrialisation it was possible for timber companies supplying sawmills to buy rights to harvest vast areas, and sometimes to possess whole villages. This situation called for state control, and it was suggested that one third of the lands allotted to farmers should be detached and made into commons. Although this idea was not initially popular amongst farmers, the first common was created in 1861. One of the ancient medieval types of common was used as an organisational model for this new type of forest commons. Over the next few decades new common forests were created, today encompassing a total forest area of about 730,000 hectares. The commons are regulated by a single law which has been virtually the same for 100 years. When the commons were first created, Sweden was an agricultural nation, and income from the commons was designated to support agriculture. Revenues from forests subsequently played a role in supporting the modernisation of farming systems. Over time, there has been a gradual shift from policies supporting agriculture to those which support forestry. Today the main part of all subsidies are designated for forestry.

THE ORGANISATION OF COMMON FORESTS

Each farmer owns a certain number of 'shares' in the common, based of the amount of arable farm land associated with the common forest. The only way an outsider can get access to a common forest is to buy, inherit, or in some way acquire a share holding farm. Currently, about 68% of the shares are in the hands of private individuals; 18% belong to companies and the remaining 14% are possessed by the church, the State and other corporate owners. A large number of commons are located in sparsely populated areas, and about 20% of the 25,099 owners can be regarded as remote owners. The general management and financial administration of the commons is the responsibility of a board, elected by shareholders. According to law, a professional forest manager must be attached to the common. This person, who is usually employed by the common, is responsible for forest management. As is the case for all forest owners, the commons are subject to control from the regional State Forestry Board. About half of the commons also run subsidiary companies or pursue commercial enterprises. The commons have developed close relations with private companies. Forest resources are managed in a sustainable manner. Only 70% of the annual increment is harvested, and the economic output

of forestry exceeds the resources that are spent to run the commons.

ACCOUNTABILITY OF DECISION MAKING

In general, organisational accountability within the commons is high. The commons hold two assembly meetings a year, open elections, and have free access to records of meetings, transparency in financial aspects, and legal rights to appeal. There are many local variations in the ways commons are organised. Two of the commons practice an administrative system retained from the first part of the 17th century. In these commons the geographical area is divided into a *rotar* with a responsible farmer in each one of them. This person is elected by the farmers. On behalf of the commons, s/he keeps track of all the changes in ownership and share holding in their rote, and also distributes the cash amounts in the area. As a consequence these commons have very good records of their owners, which is not neces-

PLATE 11: The Board of Jokkmokk Forest Common 1996.
(Photo: L. Carlsson)

74

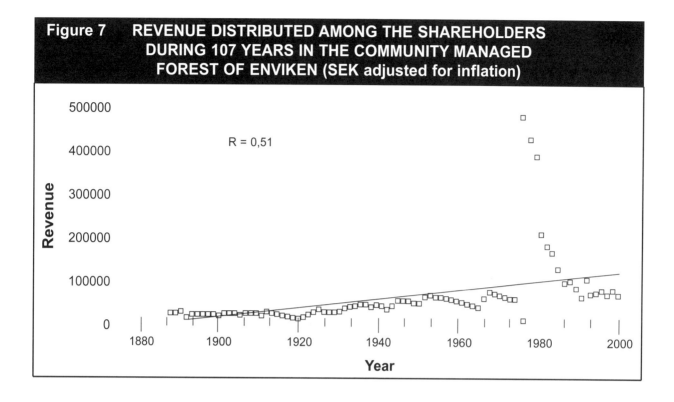

Figure 7 REVENUE DISTRIBUTED AMONG THE SHAREHOLDERS DURING 107 YEARS IN THE COMMUNITY MANAGED FOREST OF ENVIKEN (SEK adjusted for inflation)

sarily the case in other commons. One of the commons (in a high mountain area) does not have assembly meetings at all. The village elect representatives who form a mini assembly which in turn elect the board. The representatives are held responsible at local village meetings. Some commons have a significant number of elderly people, who rarely attend assembly meetings. In these commons forming groups competent to make decisions can be problematic, although this is not necessarily indicative of lack of shareholder interest. While there is diversity in local systems, the Swedish forest commons score well on accountability performance.

BENEFITS FROM THE FOREST COMMONS

The benefits for a single shareholder are threefold. First, s/he is eligible to revenue from the forest – as annual cash payments, in accordance with their number of shares. Secondly, shareholders may claim monetary subsidies given for drainage, buildings, fodder, etc. Finally, shareholders have benefits from the commons' general support to the local area, such as their provision of roads, hunting and fishing areas. About 74% of the commons distribute their residual income to

community members, usually as direct 'subsidies' to individual shareholders, for operations on their private land, such as tree planting, land drainage, etc. The commons represent both production and provision units within one organisational body, and in order to remain successful within the modern market economy they have had to pay attention to the costs related to the production and provision of these private and public goods. They have achieved this by adapting the shareholding system, keeping up with technological developments and concluding agreements with other landusers, discussed below. Figure 7 indicates that economic revenues have been evenly distributed over the years.

ADAPTING THE SHARE HOLDING SYSTEM

One problem with the Swedish common forests is the increased number of remote owners. People move from the countryside, but they tend to keep their farms and therefore their access to the commons. The additional numbers of private owners is mainly an effect of inheritance. In general the commons have adopted the principle that every farm owned by more than two persons must appoint a deputy. This person votes on behalf of the others at the assembly meetings, and is the recipient of the annual cash amounts or other types of support to the single farm. In addition, an increased number of shares are owned by forest companies, further widening access to the commons. In six of the commons companies possess more than 40% of the shares, and consequently the legal rights to a significant part of the harvest. However, in none of the commons do the companies execute their rights in proportion to their holding of shares. The election of representatives from companies to the boards is also avoided. Even where companies have representatives on the board of a common, they never push their case, and generally hold a very low profile. This situation suggests that companies have been very cautious to retain good relations with local people. If they do not, they risk problems with purchasing timber from private forest owners, the use of private logging roads and so on.

COLLABORATION WITH STATE AUTHORITIES

The commons have built alliances with several authorities involved in enforcing the Nature Conservation Act and the National Silvicultural Act, with mutual benefits. State employed and locally stationed extension rangers are responsible for all forestry related controls. The commons purchase their services for forest inventory and assessment services, and for helping control the distribution of subsidies amongst community members. By paying the state authorities for this service, the commons do not have to bear all the costs of maintaining their own control system. In addition they also protect themselves from future disputes with authorities regarding the demands for biodiversity, the preservation of protected biotopes, etc. In addition, the commons have also developed systems of co-management with local pubic institutions – schools, non-profit making organisations, etc.

KEEPING UP WITH TECHNOLOGY

When forestry was a manual enterprise, all commons had their own staff of loggers. Today there are virtually no manual loggers left in Swedish forestry. The commons have faced significant pressure to adjust to these changes. Only the largest commons have their own machinery operated by their own personnel. The largest common has 45 employees. One method of dealing with technological change is to externalise harvesting costs. Thus most commons practice stumpage sales. In this way, the buyer defrays the cost of technology, and of its improvement and renewal. Where no market for stumpage sales exists, delivery agreements and renewal felling contracts are common. These agreements can be based on harvesting with the commons' own machinery, but generally most commons have kept their machinery ownership to a minimum. The commons have also adapted in other ways through other forms of mechanisation and a reduction in personnel. The larger commons use digitised maps, computerised accounting systems, and so on.

NEGOTIATING WITH THE SAAMI

The majority of common forests are located in areas in which reindeer herding by the indigenous Saami people is practised. The Silvicultural Act # 20 stipulates that consultations must be held with the Saami before any logging can be performed on lands that they use for all-year-round grazing. The commons have negotiated with the Saami before constructing roads, harvesting, etc. Since different groups of Saami have different historical locations, and patterns of moving their herds, one basic problem for the commons is to decide which groups they would regard as 'concerned parties' or stakeholders. They have solved this problem by letting the Saami people decide themselves which groups are affected by a particular logging operation. This 'co-management' of the commons seems to function quite well. Since 1971 there has been only one appeal against a logging decision made by a common. Since the commons adjust their activities to reindeer herding, the relation with the Saami is remarkably free from conflicts.

REGULATING ACCESS TO THE COMMONS

Access questions are of considerable significance to shareholders, particularly those relating to hunting rights. Despite the trend towards rural depopulation and urbanisation, the Swedish common forests have been faced with problems of increased access for hunting and fishing. Although younger generations frequently live and work in cities, many have retained some form of ownership in the farms, many of which are now jointly owned. This enables the rights to hunting and fishing connected to the share-holding to be maintained, and enables urban members to keep social connections with their native districts. While access does not pose a problem of over-consumption, since exploitation is regulated by the state, it raises questions of who should be allowed access to the system. The commons have tried to solve this problem by creating rules to regulate hunting. For example, there are commons where only local citizens and their children are allowed to hunt.

LESSONS LEARNED

◆The Swedish common forests have survived as prosperous and competitive timber producers and providers of public goods because they have successfully integrated into industrialised society

◆The commons have adapted to changing social and economic contexts by developing systems of co-management with state agencies; adapting shareholding systems; developing systems of accountability in local decision making; keeping up with technology; regulating pressure for increased access; negotiations with other land users such as the Saami; etc.

◆The fragmentation of the centralised Swedish state, which plays multiple roles, has provided an opportunity structure which the commons have benefited from. Different units of the state have established commercial, political and juridical relations with the commons, which allow the former to promote government policies, but allow the commons to externalise the costs of controlling their own system.

CROFTER FORESTRY IN SCOTLAND: ESTABLISHING THE RIGHTS TO MANAGE WOODLANDS[4]

INTRODUCTION

This case study describes the legal and economic policy changes that have enabled crofting communities in the Highlands and Islands of Scotland to become involved in woodland and forest management. Until 1991, crofters were not legally permitted to manage woodland on their village common grazing lands, which together make up 800,000 hectares, 20% of the land area of the Highlands and Islands. After intensive lobbying, legislative changes allowed crofting communities to plant new woodlands and to protect existing ones, and to benefit from the national afforestation grants.

WHAT IS CROFTING?

Crofting is a form of land tenancy unique to Scotland, in which an individual has heritable rights to dwell on and manage a small area of land, called the croft, which is typically under 10ha. Crofts were originally subsistence holdings but have evolved into part-time agricultural units. The tenure arrangement defines a relationship between the crofter (tenant) and the owner of the land (landlord), in which both have rights and responsibilities towards each other and over the land. The arrangements were originally enshrined legally in the 1886 Crofting Act. In addition to their personal in-bye land, most crofters also have a legal share in an area of common land, called the common grazings, which is managed by an elected grazings committee, and administered by one of the committee, known as the grazings clerk. There are about one thousand common grazings across the Highlands and Islands. Typically 15-20 crofters share in an area of common grazings, on average 400-500 hectares, which is usually hill-land, unsuitable for cultivation. They have rights to graze livestock, but no rights to exploit fish, game, minerals, water or other resources of the land which belong to the landowner, and until 1991 no rights to manage woodlands. Each area of common grazings has a

Figure 8　LOCATION OF CROFTING LANDS IN SCOTLAND

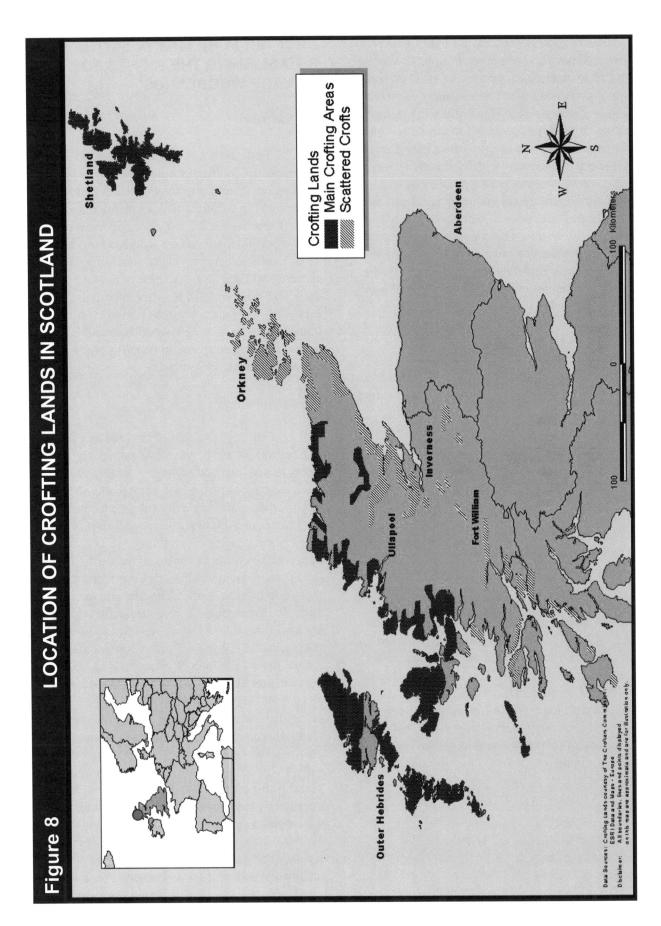

Shetland

Orkney

Outer Hebrides

Ullapool

Inverness

Fort William

Aberdeen

Crofting Lands
Main Crofting Areas
Scattered Crofts

N E S W

0 100 Kilometres
100

Data Sources: Crofting Lands courtesy of The Crofters Commission.
ESRI Data and Maps - Europe
Disclaimer: All boundaries, lines and points displayed
on this map are approximate and are for illustration only.

set of rules called the grazings regulations which regulate land use, for example, the number of stock each individual crofter may graze on the land. The overall system of crofting in Scotland is governed by legislation known as the Crofting Acts, and the system is regulated by by the Crofters Commission, a government body based in Inverness.

LAND TENURE: THE PROBLEM

During the 1980s crofters began to campaign for the right to establish and manage woodlands on the common grazings. Up to this point, the legal rights of crofters on common grazings extended only to the right to graze livestock and make some improvements to the land to aid animal husbandry, such as drainage, fencing or re-seeding. Importantly, these rights did not include the right to manage any existing woodland, nor ownership of any trees. In many areas, crofters have for generations been managing small areas of woodland, in order to provide a range of benefits including a source of fuelwood and shelter for animals. However, on the common grazings, crofters had no legal right to carry out such activities, and any trees would, in law, belong to the landowner, regardless of who planted them.

THE CASE FOR CROFTER FORESTRY

A convincing case for the benefits of crofter forestry was made, on social, economic, agricultural and not least environmental grounds. Between the two world wars, considerable areas of common grazings had been taken over by the Forestry Commission to develop plantations, with the promise of employment to crofting communities. However, with increased mechanisation and contract forestry, the number of local jobs had been in steady decline, so nationalised forestry was not delivering any social benefits to crofting communities. There was confidence that locally managed crofter forestry schemes would be more likely to deliver such social benefits as employment and training, both in the short term, from new plantings, fencing work and so on, but also for future generations of crofters.

The system of grants and subsidies in the

UK delivers significant economic benefits for managing land as woodland and it was clear that scenic and amenity benefits could also produce economic benefits through tourism, which is an important factor in the economics of crofting areas. Crofters also saw significant agricultural benefits of diversifying land use to include forestry on the grazings, including shelter, soil rehabilitation and fencing.

A notable aspect of the crofters' campaign was their emphasis on native woodland, in contrast to the predominantly non-indigenous softwood forestry which is the norm in the UK. The environmental benefits of crofter forestry include proper management of existing native woodland remnants on the grazings, many of which are ancient and of great ecological significance, and development of new native woodlands in areas which have been long deforested through neglect and intensive grazing. The knock-on effects of better protected and increased areas of native woodlands include benefits to wildlife, soils, water quality and fish resources. These benefits ensured the cooperation with the crofters campaign of vocal environmental NGOs such as the Royal Society for the Protection of Birds (RSPB), government agencies including Scottish Natural Heritage (SNH) and also with the Scottish Landowners Federation (SLF). The SLF might have seemed unlikely allies of the crofters, given their history of disagreements, and given that the crofters' aim was to increase their rights over the SLF's members' land. However, throughout most of the crofting areas, landowners interests are primarily the 'sport' on their estate, i.e.: shooting deer, and fishing salmon, and both deer and fish would clearly benefit from native woodlands.

THE PROCESS

An important lesson learned by the crofters in managing the campaign to change the law governing crofting, was that it is vital to work together. In 1985 a number of local unions of crofters, from the Western Isles, and from Assynt on the Scottish mainland, banded together to form the Scottish Crofters Union (SCU). The SCU was the most significant player in bringing about new

PLATE 12: Typical west coast sheep grazing lands in crofting areas.
(Photo: W. Ritchie)

legislation. Acting as one, the crofters were able to lobby government energetically and effectively, to employ a professional consultant to draft legislation, to build consensus amongst environmental NGOs, government agencies, and landowners, and to work with the Crofters Commission to develop a workable mechanism for implementing the new law. The House of Commons Agriculture Committee, in 1988, heard evidence from the Scottish Crofters Union, including a paper setting out the case for crofter forestry (House of Commons, Agriculture Committee 88), which led to a bill which was eventually passed by parliament in 1991.

THE LEGAL SOLUTION

The Crofter Forestry (Scotland) Act, 1991, is "An Act to extend the powers of grazings committees in relation to the use of crofting land in Scotland for forestry purposes; and to make grazings committees eligible for certain grants in respect of such use". In particular, it grants three new rights:

1. The right of any crofter to request their grazings committee to pursue forestry activity on the common grazings.

2. The right of the grazing committee, subject to the approval of the landlord, to 'plant trees on, and use as woodlands, any part of the common grazing', as long as 'not the whole of the common grazing is planted with trees and used as woodlands'.

3. The right of the grazing committee to apply for grants for woodland management and afforestation.

The former two rights involved amendments to the various Crofting (Scotland) Acts, whilst the latter involved amendments to the Forestry Act 1979 and the Farm Land and Rural Development Act 1988.

Whilst falling short of actually granting ownership of the trees to crofters, this new law enabled crofting communities to become involved in afforestation and woodland management for the

80

first time on a legal basis, and to be able to share in the benefits of these activities. The primary financial benefit was in the form of government grants, without which crofter forestry would not have been viable.

The new Scottish Parliament (formed in 1999) has promised to address the outstanding question of who actually owns trees planted by crofters on their croft lands in their proposed Land Reform Bill. Currently, legal contracts must be drawn up between landlords and crofters for each crofter forestry scheme in order to safeguard the crofters' use of their trees.

USING EXISTING INSTITUTIONAL ARRANGEMENTS

The institutional arrangements for crofting, and in particular, the grazings committees and grazings regulations, provided an established, familiar and respected framework for managing land use by communities on common land in the Highlands and Islands of Scotland. An important factor in the success of crofter forestry has been the use of these existing arrangements as the starting point to be built on when devising the legal means enabling crofting communities to plant and manage woodlands.

ECONOMIC INCENTIVES

Economic forestry requires big initial capital investment, and provides only long term financial returns. Moreover, most of the crofter forestry schemes involve native woodlands with primarily amenity and environmental benefits, which are likely to give at best modest financial returns. In Scotland, with over 300,000 each of red deer and roe deer and no predator populations, the main cost in establishing woodland is the attempt to avoid loss from deer grazing, usually by erecting 2-metre-high fences. To promote afforestation, the Forestry Commission therefore pays a substantial establishment grant, to assist with covering the costs of fencing and planting. In a crofting community, fencing and planting can mean employment, hence this grant can provide genuine economic benefits to the community in the short term.

The establishment grant is payed out in three stages: the first 50% when the initial work is completed, 30% when sufficient stocking of the ground by trees is achieved (1100 trees/ha for broadleaf species, 2000 trees/ha for conifers) and 20% five years after this. To reflect the additional costs of planting broadleaf species, and the extra environmental value of native woodland, the grants for planting native species are greater than for exotic conifer plantation establishment. Additional grants are possible for providing facilities for community access to woodland, such as footpaths, signage, or parking.

Woodland management grants are also available from the Forestry Commission. These are small annual payments on a hectarage basis, to support regular maintenance, monitoring and protection work. An additional grant, the Farm Woodland Premium, is provided by the Department of Agriculture, as an incentive for land-use change from agriculture to woodland, in order to tackle the UK's problem of unprofitable overproduction of uncompetitive food products, such as sheep. This grant is an annual income for 15 years which compensates for the loss of income from livestock removed from the land. As the heavily subsidised sheep business has collapsed during the 1990s, this grant has provided a practical way for crofters to take their sheep off the land and enable trees to grow, without loss of income. Together, these grants have acted as an incentive for crofters and others to grow trees on their land. Without them, no crofting community would have been able to make the necessary investments to begin converting grazing lands into woodlands.

One problem encountered by crofting communities, has been in the timing of the grants, particularly the initial establishment grant. All fencing work and planting must be complete before the grazings committee can apply for the grant. This can involve a significant outlay on materials and wages, and most crofting communities do not have the money available to cover these, causing significant cashflow problems. Some communities have used commercial forestry contractors to do the establishment work and carry the cashflow. This has helped solve the problem, but runs the risk that less local employment is generated although some companies are prepared to guarantee that they will offer first option of employment to local people for fencing and planting work.

EXTENSION SERVICES

Crofting communities are by nature conservative, and in most crofting townships there are few woodland management skills and rarely any knowledge of new woodland establishment. To enable them to make the land use change from grazing of stock, to woodland management, it has been essential to provide support and to inject knowledge and expertise.

The Crofters Commission has participated jointly with the agriculture, forestry and natural heritage agencies to provide the extension service necessary to enable crofting communities to get involved in woodland management. They have produced guidelines, toured the Highlands and Islands with a road-show, and dedicated a member of staff to supporting grazings committees in developing forestry plans and applying for grants, as well as managing the official approval and registry of the schemes. Full-time Native Woodland Advisors have been appointed for each region, funded jointly by the agencies, to build local capacity. Ensuring that woodland management can be integrated with crofters' other land management activities has been vital, for example, working with those with livestock to ensure fences do not unnecessarily disrupt livestock movements or shelter. The extension work of the various agencies, in demonstrating the benefits of a shift to woodland management, and assisting in the practical work of how to go about it, has undoubtedly been essential to the success of many crofter forestry schemes.

However, it is possible to have too much of a good thing! It is also important that local communities are empowered by their new knowledge, that local decision-making capacity is enhanced, and that the local people are left with a strong sense of ownership of the forestry scheme. These require sensitivity by extension workers, and in at least one of the larger crofter forestry schemes, local people feel that there has been so much involvement by such a wide range of agen-

cies and NGOs, that the result is a limited level of local control.

COALITIONS FOR CHANGE

Coalitions between a wide range of different actors have been important for securing a number of benefits to crofting communities. The most significant of these was the Scottish Crofters Union, formed by the merger of several independent unions, uniting the crofting communities. The collaboration between crofters, environmental NGOs and government conservation agencies, has helped the push for legislative change and promoted native woodland protection and development. Another example was the collaboration between several government agencies in jointly providing extension services. Finally, the collaboration between crofting communities, woodland advisors and commercial forestry managers enabled crofters to acquire the skills to establish new woods and to help overcome the problems of cash-flow.

THE RESULTS

Since 1991 85 crofter forestry schemes have been established, involving 1548 individual crofters, over 7273 hectares. The total establishment grant has been over 7 million pounds sterling, with management grants of around 0.5 million pounds sterling per year. We are now starting to see a 'copy cat' effect, where crofting communities are seeing good results from neighbouring communities and so the number of new schemes each year is on the increase. 1999 saw 31 new schemes.

LESSONS LEARNED

◆Building on the existing community land management institutions (the grazing committees) has enabled crofting communities to successfully work together to develop community woodlands and to collectively embark on land-use change.

◆The granting of security of tenure has been essential to create the long-term commitment and intergenerational security necessary for crofting communities to invest in long-term land use change, from pasture to woodland.

◆Financial incentives have been essential, and both capital (establishment) grants, and income-stream compensation were needed to encourage land-use change because the previous land-use (grazing sheep) was giving economic returns insufficient to enable crofting communities to invest in alternatives.

◆Extension services were needed to support crofting communities by demonstrating the benefits of forest management, to facilitate communities by demonstrating practical means of land-use change, and to empower communities to take full control and responsibility for the long term management of their crofter forestry schemes.

◆Coalitions and collaborations between different

groups were important in bringing about the necessary policy and legislative changes, and to provide practical support for land use change.

SAAMI REINDEER HERDERS: LOSING TRADITIONAL GRAZING RIGHTS IN SWEDISH FORESTS[5]

INTRODUCTION

The Saami are the indigenous people of Northern Scandinavia. Saami land, called Sapmi, consists of the Northern parts of Sweden, Norway, Finland and the Kola peninsula in Russia. The Saami have traditionally depended on reindeer herding for their livelihoods. In Sweden, land right conflicts and changing forestry practices threaten reindeer herding and the Saami way of life. They are lobbying for security of their usufruct rights

PLATE 13: Reindeer herds feed on lichens in the forests in the winter.
(*Photo: O. Johansson*)

83

Figure 9

LOCATION OF SAAMI LANDS IN NORTHERN SWEDEN

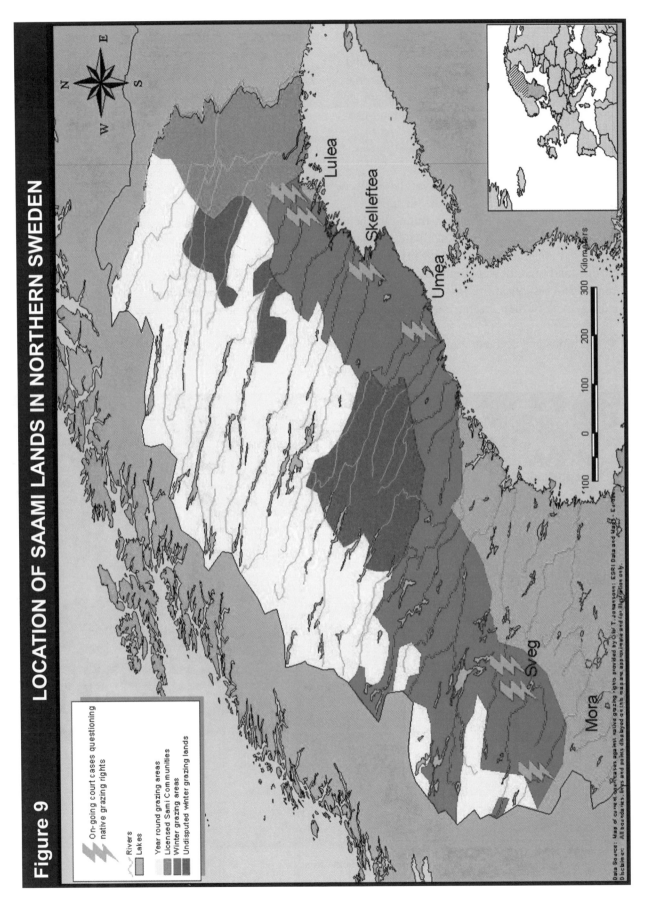

Legend:
- On-going court cases questioning native grazing rights
- Rivers
- Lakes
- Year round grazing areas
- Licensed Sami Communities
- Winter grazing areas
- Undisputed winter grazing lands

Data Source: Map of current court cases against native grazing rights provided by Öje T. Johansson; ESRI Data and Maps; Earth...
Disclaimer: All boundaries, lines and points displayed on this map are approximate and for illustration only.

and greater participation in forest management decision making, particularly decisions affecting the winter grazing of reindeer and the movement of herds.

THE SAAMI PEOPLE AND WAY OF LIFE

The Saami population numbers about 70,000 and make up part of the world's 300 million indigenous people. A Saami population lived in what is now northern Sweden before the country acquired its present state boundaries. In Sweden there are about 17,000 Saami, of which around 3000 still rely on reindeer herding for their livelihoods. Many follow a combination of occupations. The reindeer herders are organised in 52 Saami communities. A Saami community is an economic and administrative cooperative unit performing reindeer herding in a particular area. Their reindeer are semi-domesticated herds of wild reindeer that live in the northernmost parts of Europe and Asia. In 1998 there were some 228,000 reindeer in Sweden. The number varies from year to year, the allowed upper limit is 276,000.

Traditionally the Saami have led a nomadic way of life characterised by close contact with nature, making their living from hunting, fishing and reindeer herding, moving with the reindeer migration between the summer grazing lands in the mountains and the winter grazing lands in the forests. Records from 800 AD document a well established Saami reindeer herding system based on systematic use of the land. Reindeer were integral to the Saami way of life. They provided meat and milk, and every part of the animal (skins, fur, sinews) were processed into clothes and utensils. They were also used for transport. During the 20th century reindeer herding is more closely oriented towards meat production. Reindeer herding has become modernised, and today herds are followed, and moved with the help of snow mobiles, motorcycles and helicopters. Where rivers have been dammed due to hydro electric power production, lakes can no longer be used for trails, and lorries are sometimes used for transporting the reindeer to their winter grazing land. The Sami are no longer a nomadic people and have become more assimilated into the Swedish lifestyle. However, reindeer herding is a traditional way of life and part of the Saami cultural identity.

Grazing land is used collectively, and the community is jointly responsible for the tending of the reindeer. The mutual costs of the activities of a Saami community are spread according to the number of the reindeer owned. The regional authorities in each county decide on the number of reindeer allowed in each community, and this limit is conditioned by the viability of the grazing land. About 400-600 reindeer is needed for a family to live off reindeer herding only. Within any one community there are several reindeer husbandry enterprises, which may consist of one or more reindeer owners. Each owner has an independent right to make decisions about their reindeer, for example, the number of animals to slaughter.

CUSTOMARY TENURE

The land of the Saami, Sampi, was gradually colonised by Sweden, Norway, Finland and Russia during the later half of the last millennium. The Swedish parts of Sampi includes most of northern Sweden, with core areas long the mountain range. No explicit land claims have been made by the Swedish Saami, and no such rights are recognised by the Swedish state. But by using land for reindeer herding, hunting and fishing the Saami have acquired customary usufruct rights. The Swedish Reindeer Husbandry Act of 1971 recognises Saami customary rights to graze reindeer, on private as well as state land. However, while there are clear definitions of the boundaries of year-round herding areas, the legislation does not identify any clear geographical boundary for the winter grazing lands. In the case of conflict, the legislation leaves it to the Saami to prove their customary rights by written documentation in the courts. This is what is happening today.

REINDEER AND FORESTS

In summer the reindeer graze in the mountains, feeding on grass, leaves, herbs and fungi, which builds up a fat layer which helps the reindeer endure the poor winter grazing. In winter they move to forested land, where they feed on ground lichens. When lichens on the ground cannot be reached due to ice and frozen snow crusts, tree hanging lichens, found in old growth forests, are an important reserve fodder. The supply of winter grazing depends on the size of the grazing land and the availability of lichens. Grazing conditions are poor in areas which have been clear cut due to unfavourable snow quality and damaged lichens, and best in old and undisturbed forest types which provide tree hanging lichens and shelter. The best tree -hanging lichen is found in 120- 210 year old spruce forest. The lack of tree hanging lichens is a serious risk of malnutrition for the reindeer when grazing from the ground is unavailable on large areas. The reindeer does not feed on pine, spruce or any other coniferous tree.

LAND USE CONFLICTS

During the last decade, the Saami customary right to winter grazing on private lands has been challenged, primarily by private land owners and forest owner associations. The conflicts of interest that have arisen are a result of a series of circumstances for which neither party can be blamed. The root of the problem can be traced to the mid 18th century when the state first actively encouraged settlers and others to cultivate areas which the Saami had previously exclusive use of for reindeer herding, hunting and fishing. The reduced area has made it increasingly difficult for the Saami to survive from their own traditions. This has led to competition and subsequent conflicts. Part of the problem is that regulations governing reindeer husbandry have not been clearly defined.

The forest owners claim that the reindeer are causing damage to their pine plantations by scrubbing their antlers against small trees. While the Saami acknowledge that such damage does exist, they claim that it is not on a large scale and is minimal compared to the damage caused by moose. Some Saami communities have suggested that the problem could be solved by state compensation of individual land owners for the damage. No such system is at work at the moment, and the conflict has escalated and culminated in a number of legal processes. Groups of private forest owners (supported by the Forest Owner Associations and the National Farmer Organisation) are taking the Saami communities to court, questioning their right to use the land and suing them for damage that the reindeer have caused in the young trees. The land use conflicts only exist on land owned by family farms and private wood lot owners. For example, the Tassasen Saami community, from the central part of Sweden, has been sued by a group of about 50 private land owners for damage to young forests. Larger forest companies and the state fully accept reindeer herding on their land. Also, where forest management is certified according to the Forest Stewardship Council (FSC) standards, indigenous peoples' rights are recognised. The Swedish FSC standard support Saami grazing rights on traditional forest land. However, forest owned by larger companies and the smaller forest owners are not physically separated by boundaries. It is thus impossible to keep the reindeer away from private land.

COURT CASES

There are currently seven court cases, in which 12 Saami communities are involved, see Figure 9. In the absence of written documentation, which can prove long standing use of the land, the Saami are very likely to lose the court cases, followed by losing their grazing rights and way of life. Without the right to graze on private lands, the Saami communities will face problems of feeding their reindeer herds at the present number of animals, and they can not afford to pay the costs of the legal processes (which run into millions of SEK) or the compensation to the landowners. Losing the processes would basically force them to give up reindeer herding.

Some of the private land owners are prepared to accept reindeer grazing on their land if they receive compensation for the damage caused by reindeer in the young forest. The Tassasen community has asked the government to consider establishing a 'reindeer damages fund', financed by the state, and some members of the Swedish parliament have put the issue forward. A further solution would be to change the law so that those questioning winter grazing areas/rights in court have to prove that these areas have not been traditional grazing areas. In Norway the law was changed in this way when Norway ratified ILO convention 169. This means that the burden of proof lies on the side of the small forest owners and not on the Saami. The traditional land use is assumed as a Saami right. The Swedish ratification of ILO Convention 169 would help facilitate such a change.

CHANGING FORESTRY PRACTICES

Since the 1960s there has been an increase in clear felling; the establishment of monocultural plantations; cutting of old growth stands; road building, and other industrial techniques associated with commercial forestry practices. Changing management have threatened both reindeer winter grazing and biological diversity. Large areas of old growth spruce forest are being clear cut and replaced with planted trees. New plantations are being felled before they have aged sufficiently to provide the optimal environment for tree lichens. Since tree lichens spread from older to younger trees, large clear cuttings and young forest areas without older trees impede the dispersion of tree lichens.

The Saami are concerned that commercial practices reduce the availability of winter fodder for the reindeer, and constrain the movement of herds. They are calling for greater participation in decision making in forest management practices which affect the availability of their winter grazing and movement of reindeer herds. In particular, they are calling for limits to clear cutting, which can affect all year round access to contiguous grazing lands; migration trails; resting pastures, and facilities that have traditionally been used for reindeer herding. They want a percentage of lichen bearing trees to be left standing for a longer period of time to ensure adequate dispersal of tree lichens in commercial forests. And they want more participation in the planning of logging roads to ensure they do not divide reindeer pastures and make it difficult to keep the herd together during grazing. These demands are supported in the handbook of Swedish Forestry law, but multiple and co-use of the forest has proved difficult to implement in practice. The Saami have been forced to tolerate considerable encroachments on their reindeer herding rights.

DISSEMINATING INFORMATION

In recent years the Saami have been more proactive in promoting their interests, by disseminating information and presenting alternative ways of resolving conflicts. Since early 1999 information has been disseminated internationally to a broad range of interest groups, including NGOs, the general public, politicians, mass media, the Swedish forest industry and their European buyers. The Saami have also initiated dialogue between different stakeholders such as NGOs, forest industries, timber buyers and politicians. As a result, the Saami case has become better known nationally as well as internationally by the general public and within political and forest management bodies. Interest in their cause and commitment to helping resolve the conflicts have increased significantly.

NETWORKS AND COALITIONS

Saami communities have also been proactive in building and collaborating with a European network of environmental and social NGOs with a common interest in defending ecologically and socially sustainable forest management, particularly those committed to respecting indigenous peoples' traditional rights, multiple forest use and conservation of old growth forests. For example, active allies of the Tassassen Saami community are the 11 other Saami communities defending their traditional rights in court; national Saami organisations like the National Swedish Saami Association (SSR), the Saami Parliament and the International Saami Council. Outside Saami circles there is an increasing number of European Environmental NGOs (Greenpeace, WWF, Friends of the Earth, Robin Wood) and indigenous peoples support groups (e.g. IWGIA; Society for Threatened Peoples) as well as smaller individual organisations supporting indigenous peoples' rights. International networking organisations like the Taiga Rescue Network (TRN) and the World Rainforest Movement are supportive as well. Cooperation and dialogue with the FSC and Swedish forest companies, and buyers of Swedish wood products, is ongoing and encouraged. The network of NGOs facilitates international information dissemination through all available channels and ensures Saami representation in relevant discussions. Media work and cooperation with these networks and coalitions is ongoing.

◆Multiple-use of the forest by forest owners and reindeer herders is possible in principle and does work in many areas. However, co-use by the Saami and small forest owners has led to conflicts in some areas of northern Sweden. An increasing number of forest owners, backed by their associations and hunter associations, are actively going against the Saami by initiating court cases and suing them.

◆Present legislation in Sweden does not guarantee Saami access to their traditional winter grazing land. The ability of the Saami to have their land claims examined in court is largely theoretical. They do not have the financial resources to pay legal costs. Furthermore, the written documentation of traditional usufruct rights demanded by the courts does not exist.

◆The adoption of modern forestry practices, such as clear cutting, establishment of monocultures, loss of old-growth forest, has been detrimental to both reindeer herding and biodiversity. Such practices reduce the availability of hanging lichens which provide winter reindeer fodder. These lichens are found in forests of high conservation value.

◆Coalitions with human rights and environmental NGOs has been beneficial to Saami by providing channels through which they can promote Saami interests and disseminate information nationally and internationally. International support is very much needed in order to give the Saami a voice in their own country. Collaboration with forest industry and independent forest management bodies, such as the FSC, which support indigenous peoples' rights for sustainable forestry, has also been beneficial to the Saami.

◆Several solutions to the present land use conflicts have been identified. One is to develop new maps which indicate the boundaries of traditional winter grazing rights. This approach has been adopted by the FSC in Sweden. A further solution would be to change the law so that those questioning winter grazing areas/rights in court have to prove that these areas have not been traditional grazing areas, as adopted in Norway. The Swedish ratification of ILO Convention 169 would help facilitate such a change.

◆Other solutions include Government financial support to Saami in court cases; the establishment of a reindeer damage fund; the promotion of certification schemes which adopt at least minimum standards for co-use of forest and for reindeer herders rights; and greater participation in forest management decisions, particularly those which affect winter grazing and movement of herds.

COMMUNITY FORESTRY IN THE BORDERS REGION OF SCOTLAND: CONTRIBUTING TO NATIVE WOODLAND RESTORATION[6]

The Borders Forest Trust (BFT) was established in 1996 through grassroots community action and popular support for the restoration and expansion of native woodlands in the Borders region of Scotland. The Borders region has lost more than 95% of its native forest cover, more than any other area in Scotland, as a result of millennia of forest clearance for agriculture, predominantly sheep farming and arable crops. As native woodland has declined, so the level of associated employment and the significance of timber in the local economy has declined. The BFT believes that community and native woodlands can be managed as a valuable economic, educational and recreational resource, as well as providing important habitat for a broad range of native plants and animals and valuable landscape features.

The BFT is a not for profit charitable, membership organisation. It has a community-based approach, involving local people in the long-term management of newly established and mature woodlands, while promoting community ownership (either legal or managerial), and access to all. The Trust also works in partnership with a wide range of statutory and voluntary organisations, and acts as a regional platform for bringing

PLATE 14: Artistic events in Wooplaw Wood, UK.
(Photo: Borders Forest Trust)

together groups and individuals interested in the
conservation and expansion of native woodlands
to deliver collective aims and objectives effectively. In 1998 over 90% of BFT's income of
£460,705 (pounds sterling) was in donations and
grants from a wide range of bodies including the
Millennium Forest for Scotland Trust, Scottish
Natural Heritage, Forestry Commission, European
Funds, and WWF.[7] In recent years a funding
strategy has been developed to help BFT become
less grant dependent, and to increase the
percentage of its income from private donations,
fundraising appeals, and the sale of goods and
services (timber products and specialist advice).
Many of the BFT projects combine social, environmental and economic objectives and benefits,
as the examples indicate below.[8]

WOOPLAW COMMUNITY WOODLAND

Wooplaw Wood was bought as a commu-
nity resource in 1987, and is reputedly Scotland's
oldest community owned woodland. It is located
in Selkirkshire some 35 km South of Edinburgh

near Galashiels. It is approximately 25 hectares in
size. Membership is open to anyone whether
living near or far. The woodland was originally the
inspiration of wood sculptor Tim Stead who lived
in the nearby village of Blainslie. In the early
eighties Stead increasingly felt the need to plant
trees such as the native elm (*Ulmus* spp.), ash
(*Fraxinus excelsior*) and oak (*Quercus* spp.) to
replace those that he had used in his furniture
making and wood sculpture. In 1986 he initiated a
fund-raising programme to buy land a community
woodland. When Wooplaw Wood came on the
market in 1987, Borders Community Woodland
managed to raise the asking price of £ 33,000
(sterling) in six weeks. There was considerable
grassroots support for the purchase with donations
from the local and wider community. Some three
hundred people joined the charity 'Borders
Community Woodland', paying a membership fee
of five pounds. A substantial contribution was also
made by WWF Scotland, and the Countryside
Commission for Scotland.

According to Stead[9] the idea of the

Border's community wood is to "involve people, give them access and give them pleasure". It is for people who care about woods, and a place where the diverse aspirations of a community may be realised. He believed that people from surrounding areas should be able to drive, walk or cycle to the wood and get involved in various activities including the planting of trees and maintenance operations, as well as picnicking, walking and simply enjoying their surroundings. In the late 1980s the wood was under utilised by the local community, even though it is situated only a few miles away from a number of villages and towns. Stead claimed that "..it's an idea ahead of its time".[10]

However, since the mid nineties the management of the wood has become the focus of numerous training, cultural and artistic events. In addition to providing access to local communities, the objectives of the woodland are to enhance wildlife habitats, and re-establish areas of native woodland. A further objective is to manage the area for timber extraction, creating revenue that can be channelled back to management of the woodland.[11] The woodland is popular with various organisations who practice woodland and conservation skills, and between three and four hundred people are involved in such activities during the year. Much of the work to date has been carried out in partnership with the Scottish Wildlife Trust (SWT), who use the wood for training e.g. tree felling, chainsaw work, fence construction, drystone dyking, path-making etc. The wood contains wildlife such as great spotted woodpeckers (*Dendrocopos major*), tree creepers (*Certhia familiaris*), woodcock (*Scolopax rusticola*), roe deer (*Capreolus capreolus*) and badgers (*Meles taxus*), and is used by wildlife enthusiasts, local RSPB activities and for school educational excursions. The woodland may also provide casual employment and income for one or two people during planting, weeding, felling and pruning activities. Some woodland income is obtained from thinnings and brash, which may be used for charcoal making. Artistic events, barbecues and *ceilidhs* (traditional music and dancing) also attract many people into the woods. A recent *Woodland Access Initiative* has allowed even more groups to participate in woodland activities, by providing improved path and boardwalks, interpretative maps and trails, outdoor classroom and educational facilities. Hundreds of trees have been planted by volunteers as part of a carbon sequestration project. Decision-making is guided by a woodland warden committee and carried out by eighteen community-woodland wardens. While local people are given free access and considerable freedom to use the wood as they wish, the potential dilemma of any one individual or group dominating the resource at the expense of others has never arisen.

WOODSCHOOL[12]

The Woodschool initiative, at Monteviot Nurseries near Jedburgh, was established within BFT in 1996. Its objectives are to increase the amount of low to mid-grade timber that is processed locally, and to bring creative designers and furniture makers into contact with the under-valued and underused hardwood resource. Like other rural areas in Europe, the hardwood timber market and product demand in the Borders region is constantly changing. Prices, volume, species, quality, dimensions and products are amongst the variables affecting this shifting scene. It is estimated that some 90% of the hardwoods felled in the Border's region is exported for conversion in the North or Central England, and brought back to Scotland as boards or finished goods. This creates an economic and ecological imbalance in the region. A greater local return can be achieved by developing high value timber products and by encouraging local people to engage in economic activities associated with a woodland culture.

Woodschool currently offers bench space to some 8 designers and furniture makers for a monthly bench (rental) fee. It provides extensive workshop and design studio facilities and marketing support. The resident makers have designed and produced items for restaurants, offices, shops, as well as prototyping for architects and interior designers, and installing work in school playgrounds and other public areas, and working on private commissions. Examples of high profile work include the Icelandic Parliament Speaker's Chair, the installation of choir seating in St Mary's Cathedral, Glasgow and public seating in the National Museums for Scotland. Woodschool also offers courses in green woodworking and wood turning to local people. Woodschool's business development plan is enabling it to reduce its dependence on grants and become financially independent.

◆Native woodland restoration and management brings real tangible social, economic and environmental benefits to local communities.

◆The vision and support of local leaders within the local community have underpinned the development of community-based forest initiatives in the Borders region.

◆BFT has learned that a strong partnership approach brings great strength and breadth of knowledge to the development of the projects. Project partners and stakeholders need to be involved in projects at the start of the planning process. The success of BFT's work is based on the enthusiasm and ability of an enormous range of individuals, communities, agencies and organisations.

◆Community led projects are more likely to result in sustained activity, although community participation takes time and often results in changes in project plans. The project development process is as important as the actual project outcomes in terms of community development. Projects need continual evaluation and revision in response to changing contexts.

◆One of the major lessons learned over the last five years by BFT is that most environmental and community NGO's do not have the capacity to deliver environmental/community value and efficient project and financial management. The BFT has relied on start-up grants and donations from a wide range of bodies, highlighting the importance of material support, and economic incentives for the development of community-based projects. However, in order to achieve economic sustainability, the BFT recognises that it must reduce dependence on external grants and increase the proportion of its income derived from membership, fundraising and marketing its

own goods and services. The future corporate strategy, structure and management of BFT will be analogous to a small-medium sized enterprise/business.

COMMUNITIES AND BIODIVERSITY CONSERVATION IN THE MEDITERRANEAN REGION: THE UNDER-UTILISED POTENTIAL OF NON-TIMBER FOREST PRODUCTS[13]

INTRODUCTION

Non-timber forest products (NFTPs)[14] have held an important role in the economies of Mediterranean cultures since prehistoric times. NTFP production is still an important aspect of forest management, often contributing more to economies than timber. Unfortunately, demographic and economic changes during the last hundred years have resulted in a diminished role for the production of NTFPs. The effects of these changes have been in many cases detrimental both to biodiversity and community welfare. However, with shifts in forest management thinking, careful ecological research, and strategic marketing, many NTFPs have potential to enter the organised market and bring benefits to producer communities.

WOODLAND MANAGEMENT SYSTEMS[15] AND NTFPs

NTFP systems play a significant role in household economies and food security as well as national economies, and contribute to landscape value, tourism potential, cultural integrity and are compatible with environmental objectives. The continued importance of Mediterranean NTFPs in rural economies has preserved distinctive forest structures that, even though anthropogenic, hold significant biodiversity. The traditional and familiar Mediterranean landscape of woodlands alternating with pastures and cultivated lands (agro-sylvo-pastoral systems) is a result of this type of sustainable rural management. NTFPs are the output of this management. Mediterranean NTFPs are numerous, and include products given in the table below.

Table 9 — NON-TIMBER FOREST PRODUCTS IN THE MEDITERRANEAN

Tree Products	Examples	Region/countries
Bark	Cork (*Quercus suber*)	Spain, Portugal, Italy, Tunisia, Algeria, Morocco
Resins	Mastic gum (*Pistacia lentiscus*) Styrax oil (*Liquidambar orientalis*)	Greece Turkey
Fruits	Chestnuts (*Castanea sativa*) Walnuts (*Juglans regia*) Olives (*Olea europaea*) Carobs (*Ceratonia siliqua*) Figs (*Ficus carica*) Pine nuts (*Pinus pinea*)	All over Northern Med. All over Northern Med. All All All Portugal, Spain, Italy, Turkey, Lebanon
Leaves	Laurel leaves (*Laurus nobilis*)	All
Firewood and charcoal		All
Others (understorey) **Medicinal, aromatic** **and edible plants**	Liquorice (*Glycyrrhiza glabra*) Arnica (*Arnica* spp.) Thyme (*Thymus* spp.) Oregano (*Origanum* spp.) Lavender (*Lavandula* spp.)	Northern Mediterranean Northern Mediterranean Most countries Most countries Most countries
Wild flowers		
Mycological products	Mushrooms, truffles	Most countries
Apicultural products	Honey, propolis	All

Multiple purpose forest management may combine several activities such as resin collection, apiculture, herbaceous plant gathering (from the understorey), animal husbandry or employment in recreation.

SOCIO-ECONOMIC BENEFITS FROM NTFPS IN THE MEDITERRANEAN

There is a great economic incentive for countries in the Mediterranean to develop the NTFP production potential of their forests and generate positive socio-economic benefits for rural populations that are compatible with conservation values. However at present, the production of NTFPs in the Mediterranean is highly neglected, with the possible exception of resins and cork.

NTFP production has always played a significant role in the welfare of forest communities, many of whom live marginally at the edge of subsistence. In some low-income areas of the Mediterranean, NTFP production (such as resin collection) was – and continues to be – the only reliable source of labour and income. In addition to tangible, measurable socio-economic benefits from enhancement of NTFP production, NTFPs also have a cultural significance for many rural societies. Preservation of forest management methods will not only preserve environmental and socio-economic integrity but will also permit cultural continuity.

The actual production of NTFPs that takes

place in Mediterranean forests is much less than their production potential. Table 10 gives an overview of the actual production for selected NTFPs compared with the potential production in tonnes per year.

Another indicator is the NTFP production potential in terms of monetary units coming from Mediterranean forests.

Comprehensive statistical data on production and trade of NTFPs are needed for an accurate estimation of their true socio-economic contribution to sustainable development. Such information would help in the elaboration of appropriate policies for NTFP production and promotion. Such policies, if accepted by senior decision makers could lead to a more equitable access to NTFP resources and to a fair distribution of benefits obtained by local communities.

Many factors have hindered the commercial development of NTFPs, the principal ones being:

◆lack of appreciation for the economic potential of NTFPs

◆ignorance of the importance of NTFPs in rural societies

◆bias against the quality of non-domesticated resources

◆predisposition of agricultural professionals towards products that need higher technological inputs and complicated processing techniques

◆a general lack of knowledge on NTFPs

◆substitution of many natural substances by synthetics

There are many reasons behind these hindering factors. Since NTFPs are used locally for subsistence or sold in rural markets, information on their market value has been excluded from official statistical data. The division of official governmental agencies between forestry and agriculture creates some practical inconvenience for the consideration of NTFP commodities that fall within a grey production area for many experts. Finally, the mainstream forestry view that the value of forest resources clearly reflects their timber production potential, has resulted in only incidental and fragmented consideration of NTFPs.

NTFPs AND MEDITERRANEAN FOREST CONSERVATION

The role of NTFP production in forest conservation in the Mediterranean is mixed. On the one hand, problems have arisen from rural abandonment, and the decline of multiple management practices in forests that produced NTFPs. For example, in the past thriving rural populations managed the forests on which they depended, resulting in less frequent fires than nowadays.

Table 10	ACTUAL AND POTENTIAL PRODUCTION OF SELECTED NTFPs	
NTFP	**Actual Production (tonnes/year)**	**Potential Production (tonnes/year)**
cork	3,750,000	9,135,000
sylvo-pastoral	2,021,684	6,458,081
wild fauna	1,211,527	3,873,632
medicinal & aromatic plants	4,546,965	14,528,120
mushrooms	757,827	2,420,967

On the other hand, many problems have arisen with the intensification of NTFP gathering and production, both in the northern and southern Mediterranean. In recent years, the increase in non-local and foreign market demand has resulted in unsustainable gathering of many NTFPs, mainly medicinal plants, herbs and bulbs, or other produce such as pine nuts. Additional problems have arisen with the development of foreign production centres and of chemical substitutes that have contributed to the collapse of products such as resin. Genetic decline is also an emerging issue: the destruction of natural habitats and the introduction of modern cultivation which neglects local varieties of some species have resulted in narrowing down the genetic diversity of both wild and cultivated types of the species.

In the past, traditional NTFP production in Mediterranean woodlands maintained a delicate balance between human presence and forest integrity. However, in order to integrate active NTFP production within current conservation efforts in Mediterranean forests, there are many questions related to NTFP ecology and production that need to be answered.

COLLABORATION OF DIFFERENT STAKEHOLDERS

Actions and future development needs to be undertaken by coalitions of experts, including representatives from local communities, non-governmental and governmental organisations, which will follow a multidisciplinary approach to the issue. Only a holistic, interdisciplinary approach could bring success to any commercialisation attempt for NTFPs. The WWF Mediterranean Programme currently targets the conservation of significant forest areas through the promotion of the sustainable production of NTFPs, with its consequent economic benefits for rural communities, and is building partnerships with a number of Mediterranean environmental groups and institutions.

COMMUNITY INVOLVEMENT IN CHESTNUT PRODUCTION IN THE PARNON AREA, GREECE

The Parnon mountain range lies in the eastern part of Peloponnese in southern Greece. It is covered by forests of Greek fir *(Abies cephalonica)* and black pine *(Pinus nigra)*, juniper species, including *Juniperus drupacea*, a relic species with a European occurrence only in Parnon. The forests also have a rich bird and reptile fauna, and are of significant conservation value. Chestnut *(Castanea sativa)* woodlands occur naturally in Parnon, but are also planted by local populations. There are 809 hectares of chestnut groves in the region. The woodland of Castanitsa has a community property status, and belongs to the dwellers of rural communities as a whole. Chestnut producers have organized themselves in a Co-operative. Average chestnut production is about 250 tonnes per year. About 1/3 of the woodland has been given organic certification. There have been no fires within the chestnut woodlands since 1966. This is ascribed to the fact that human activities within the woodlands, such as chestnut harvesting, results in good forest management. Until recently, chestnut production

Table 11	US$ PER HECTARE FOR SELECTED NTFPs[16]
NTFP	**$US per hectare**
cork	264,600
sylvo-pastoral	560,280
wild fauna	403,200
medicinal & aromatic plants	252,000
honey	1,915,200
mushrooms	1,764,000

in Parnon was characterised by sporadic, unorganised production with no management plan, and lack of market promotion. The co-operative is developing appropriate management practices for all stages of production and a dynamic strategy to promote chestnuts in the market.

Chestnut propagation is done by grafting and seeds. Fruit production starts when the tree reaches 5 years of age. Local varieties are edible, but cannot compete with Italian or French varieties on the market. Fruit morphology limitations make Parnon chestnuts more difficult to process. As a result local Parnon farmers often graft chestnut trees with foreign maron varieties, which have better confectionery properties, but may threaten local natural genetic diversity. The *Cryptonectria (Endothia) parasitica* can cause severe damage especially in grafted trees.

Current plans for Parnon focus on the establishment of a region-wide Parnon National Park, which shall include the coastal wetland of Moustos, one of the most important wetlands along the eastern shores of the Greek – mainland. Representatives of local communities supported the initiative for the establishment of the protected area. A management plan for the Park, which involves chestnut production, has been prepared. A training programme addressing the needs of chestnut producers – has been completed, and the WWF Mediterranean Programme had an active role in this training. The programme exposed producers to chestnut cultivation, organic production methods, NTFP certification and marketing promotion techniques. Chestnut production can play a key role in the conservation and development of the region.

LESSONS LEARNED

◆NFTP production has the potential to provide conservation and community livelihood benefits in the Mediterranean region.

◆The time is ripe to act and establish small to medium scale operations, which would organise forest management for NTFPs, address ecological and social needs of rural populations and preserve the centuries-old rich natural and cultural landscape.

◆Investments need to locate and capitalise on existing knowledge of multi-management of forests. Any aid given for the introduction of new economic activities should be tailored to the existing potential of Mediterranean woodland resources.

◆Future developments should be based on the participation of multiple stakeholders, including representatives from local communities, non-governmental and governmental organisations, who can pursue a multidisciplinary approach to NTFP issues.

URBAN COMMUNITY FORESTRY IN LONDON: THE ROLE OF SMALL SCALE WOOD PROCESSING[17]

INTRODUCTION

The Urban Community Forestry Project was established in 1996 by the BioRegional Development Group (BDG), a local environmental NGO based in the London Boroughs of Croydon and Sutton. The project works to involve urban communities in combating two problems particular to the urban environment in the UK: the waste of the by-products of tree surgery and the neglect of urban woodlands. By addressing and integrating the economic, social, and ecological components of sustainability, it seeks to be a working example of sustainability in urban areas, and to inspire others to adopt good practices elsewhere.

BACKGROUND

In London it is estimated that about 18,000 tonnes of tree surgery material are generated per annum. Some 51% is sent to land-fill sites and 11% is burned. This is a significant waste of a valuable resource. The majority of London's woodlands are also under-managed. This is especially serious for ancient-semi natural woodlands where the decline of active management has reduced diversity and resulted in the decline of many species. Urban woodlands have also suffered from the lack of appropriate care because the culture of tree care in cities has been one of arboriculture not forestry. This has led to both inappropriate and expensive intervention in London's woods, and the view that the care of trees in the urban environment is a cost item, not the management of a potentially valuable resource. The BDG is attempting to bring management back to neglected woodlands, and to stimulate the local urban economy and employment by supplying products on a local basis. It has helped establish a national market for two bulk woodland products, charcoal and firewood, and thus help replace unsustainable charcoal imports.

A 'TREE STATION'

In order to give value to a product often regarded as waste it is necessary to develop an infrastructure and methodology to accumulate material and create saleable products. The solution developed by the BDG was the Tree Station, a site that can agglomerate arboricultural arisings (tree surgery waste) from street trees, parks and urban woodlands, split them into various fractions, which can be made into salable products, or collected and sold on. The layout, plant and methodology of a Tree Station are influenced by the product type, market and ease of access.

With the exception of tree butts of great value, the cost of transporting timber from woodlands is often greater than its retail value. Timber felled in these circumstances is often left in situ, burned or cut up for disposal as waste which is a sad end for a potentially valuable resource. This is often the case in small woodlands or parks and for street trees where the amount of timber generated is too small. In this context, a tree station has value in that it can agglomerate timber from a variety of local sources. The wood received from arboriculture is varied and typically comes in the form of rings or cordwood cut at random lengths and sizes usually small enough to be lifted with ease onto a contractors pick up. This renders it unsuitable for the pulp market which demands reasonably straight logs of between 1.8 and 2.3m long and is the normal outlet for low value timber which used to make paper or board. However, this material is ideal for lump wood charcoal sold to the barbecue market.

Arboricultural 'arisings' will not be

removed from the waste stream if the products cannot be sold. It is therefore vital that markets and the infrastructure to supply them are established first. The BDG have developed markets with the large retailers such as B&Q (a large DIY store) and BP petrol station forecourts. These are supplied via a network production system of local charcoal producers throughout the country. Much of the timber in arboricultural arisings is also suitable for firewood which is sold on a local basis either bulk or via hardware shops and petrol station forecourts.

To date some 800 tonnes of tree surgery material have been removed from the waste stream and turned into products which have been sold locally. The income generated by the tree station is enough to cover the wages of two full time employees, and some capital costs. The work in 1999 generated some £30,000 (pounds sterling). The tree station puts money back in a tree resource area, which stimulates further management and community involvement. The supply of locally produced charcoal by the BDG helps reduce foreign charcoal imports often sourced from unsustainably managed environments such as mangrove swamps in south-east Asia.

WOODLAND MANAGEMENT

Croydon contains some of London's largest ancient semi-natural woodlands. Historically actively managed these have been largely neglected especially over the last fifty years and have lost much of their structural diversity with a consequent negative impact on certain flora and fauna. Any work being undertaken in the woods was reactive and carried out for health and safety reasons. The London Borough of Croydon was charged for work done on a per tree basis, and any timber generated was burnt or left in situ. Tree management rarely altered a woodland's structure. This approach also meant that Croydon did not take advantage of forestry grants, and rarely looked at the woodlands with any long term plans. This further discouraged positive intervention.

This trend was reversed by the BDG who, in partnership with the Council and local commu-

PLATE 15: Croydon Tree Station
(Photo: J-P. Jeanrenaud)

nities, have put in long term management plans
and made use of Forestry Commission grants. This
has enabled the revival of forestry management
over some 160 hectares of woodland with costs
being met by sale of timber and grant monies.
Results are already being seen in the compartments put back into management. In Kingswood
orchids have returned after several years' absence,
and flowers have also returned in managed blocks
in Threehalfpenny Wood, unlike neighbouring
neglected blocks.

The Project's role in Sutton has differed
from that in Croydon. With a small less varied
woodland cover, the emphasis has been on
increasing tree cover. A new urban woodland has
been planted on an area of disused allotments.
Here, one acre of allotment was planted under a
WGS as a mixed broadleaved wood with a
predominantly hazel understorey. The intention
here was to create a valuable new environment,

and to provide a range of products that are locally
useful. For example, hazel rods can be substituted
on neighbouring allotments for imported bamboo.
The woodland has now been planted and is being
maintained and once established will be provided
with an interpretation board.

COMMUNITY INVOLVEMENT

The BDG views partnerships with active
community groups, and local government authorities as crucial for local woodland and tree management. In Kingswood BDG works with the 'Friends
of Kingswood'. This is a community group of
some thirty members, of both genders and from a
variety of socio-economic backgrounds, who are
highly motivated by ecological interests. The
group has a formal structure including a chairman
and committee, and is supported by a grant from
the London Borough of Croydon. Woodland work,
such as the preparation of management plans,

coppicing and ecological surveys and monitoring are carried out in peoples' spare time. Some of this work has benefited from Forestry Commission grants. The BDG has helped develop markets for timber for local groups and encouraged their active involvement in the development and execution of management plans. Elsewhere, the project has helped establish charcoal making facilities at a local centre of people with learning difficulties. They have their own mini tree station receiving tree surgery wood and turning it into charcoal which they sell at their craft centre. BDG regional staff on this project are usually drawn from the long term unemployed and are trained as appropriate. The London Boroughs of Croydon and Sutton have always been leaders as regards environmental policy and practice, and have been at the fore as regards developing Agenda 21 and environmental policies and initiatives. Without their proactive approach, the project and partnerships would not exist.

EDUCATION

The decline in management of urban woodlands is largely a result of the lack of demand for its products. With little activity in urban woodlands, and little need for the products they used to provide, the wider public tend to view active management negatively. The project has sought to address this by a wide ranging education campaign ranging from putting up notices with detailed explanations of work in progress, to lectures and school visits. An essential part of the Bioregional Development Group's education programme is getting local wood products into the community and letting the products themselves tell the story.

CERTIFICATION

The BDG have obtained Forest Stewardship Council (FSC) certification for the management of the urban tree resources in the London Borough of Croydon. This is the first time, internationally, that highway trees have ever been certified according to FSC standards. The BDG believe that the management of urban tree resources should be informed by a sustainable, holistic and long term perspective, and be seen as an integral

component of the world's timber resource. The urban timber resource can be viewed as a plantation, with the majority of trees in cities and towns having been planted, save the occasional remnants of natural forests. If managed efficiently, products from urban tree resources, like plantations, can play their part in reducing pressures on endangered natural forests.

Currently the main system for achieving a sustainable tree management is via certification to the FSC standards, which have become international benchmarks of good practice. The growing acceptance of these standards has stimulated the development of the UK Woodland Assurance Scheme (UKWAS) to accredit its own woodlands and form the standard for others. Certification to FSC standards has several benefits. It requires the establishment of good working management systems which are regularly monitored. It allows the sale of products to timber buyer groups committed to purchasing timber and wood products from certified sources. This helps open markets for woodlands and their products. This 'international first' in certification of urban trees is a vital step for acknowledging that street trees are an important resource and can be managed sustainably.

LESSONS LEARNT

◆Partnerships between local urban communities, the BDG (the local NGO) and the local government authorities have been essential for the development of the Urban Community Forestry Project. The partnership has demonstrated that trees in the urban environment are a resource not a burden, and helped demonstrate that the sustainable ideal of 'local products for local needs' is viable.

◆Supplying mainstream retailers with tree and woodland products, such as charcoal, has been vital because they provide a volume market. This has enabled significant amounts of material to be taken out of the waste stream and an appropriate level of management to be introduced into local woodlands.

◆The availability of Forestry Commission grants has been a key incentive for participating groups to undertake tree and woodland management. Communities have also benefited directly from the sale of products. Involving the community at a real level – where the fruits of their labours do go into products – has helped sustain the partnerships.

◆ Maintaining flexibility within each community group, and woodland management approach has been vital for the development of the Urban Forestry Project. Each group has its own character and wants to tackle different aspects of woodland management in its own way. Although in close proximity, each wood behaves differently when brought back into management.

◆ Adopting new small scale technologies, such as the tree station, has been key for improving product processing and woodland management. While woodland management and charcoal making bear the burden of long traditions, it is always worth considering that old ways are not necessarily the right ways and that alternative methods of silvilcultural management and more efficient charcoal production can provide a more sustainable future.

FOREST WORKERS IN EUROPE: THE ROLE OF UNIONS IN DECISION MAKING[18]

FOREST WORKERS IN THE COMMUNITY

Forest workers are often long established members of rural communities, and in some areas there are long family traditions of forest management work, such as woodcutting. Forest workers are also key stakeholders in forest management processes, and are involved in implementing forest management decisions on the ground. They frequently have good insights into sustainable forest management management. At a local level, they may be regular or contract employees, consumers of forest products, and users of other forest leisure and environmental services. As community members, the wages they receive from forest work are then passed though the community leading to secondary (typically service) jobs within the community. But woodworker perspectives and interests are often neglected or inade-

quately represented in forest management decision making from both local to international levels. Wood worker unions have developed throughout Europe during the last 100 years to promote woodworker interests and greater worker involvement in forest management decision making.

FOREST WORKER UNIONS: A COMMUNITY OF INTERESTS

Many forest workers within Europe are members of national unions.[19] Unions support their members through negotiations, campaigns and solidarity actions to protect wages and working conditions. Core requirements for unions are safe, stable and well paid jobs. In spite of the progress of technology, forestry work remains one of the most dangerous occupations in most countries. Only a few forest workers reach the normal retirement age without occupational diseases or physical deterioration. The negotiation of collective interests can be considered a special type of community involvement in forest management – a community bound by interests, rather than by locale.

Today, traditional union issues have widened to include more fields of concerns. In forestry this includes active participation in sustainable forest management. This has occurred because unions, like other stakeholders, recognize that there are no jobs in the forest and timber sector if there are no forest resources. Union leadership has recognized that full and active participation in 'new issues' is closely linked to increased responsibilities, and as such it requires increased commitments. The increased commitment results not only in a more motivated staff but it also increases its knowledge and ability to engage with the other stakeholders at a community and informal level, as well as at the professional level.

Four major federations exist for forest workers in Europe:

1) Nordic Federation of Building and Wood Workers

2) European federation of Building and Wood Workers

3) International Federation of Building and Wood Workers

4) European Federation of Agriculture.

UNION INVOLVEMENT IN DECISION MAKING

Participation is a basic need for workers in the forestry sector. Unions have a long tradition of developing their own models of involvement. Unions are present at all levels of forestry work: international, national and forest management unit levels. Union involvement has to be understood in different ways, as workers may be state employees, employees of companies, self-employed, or contract workers, with different participatory roles. To bring safety and health into forestry work and to secure the forest resource, all expert levels should be used. This includes the manager at the political level, the union at the enterprise level and the workers themselves at the work-site. For unions, 'partnership' could be a possible positive outcome of a participation process with equally distributed rights and duties for each participating person or group. There are many examples of unions' involvement in decision making:

◆*Participation with non-governmental organizations and local communities* – Unions are forming alliances with international NGOs in the forestry and environment sector – notably in establishing eco-labelling guidelines and socially and environmentally friendly forest restoration projects and energy saving devices. At local levels they are also involved in community based projects, such as afforestation projects.

◆*Participation at the Enterprise Level: Collective Agreements* – Collective bargaining is, even in countries with relatively high wages, a tough, but daily task for union leaders and workers representatives at the enterprise level. Collective agreements are typically found at the level of larger enterprises with long-term employees. Unions are also trying to extend the concepts of collective agreements to contract work. For example, in December 1999, following the heavy storm *Lothar*, unions negotiated with German authorities to obtain hourly wages with additional premiums for high quality work in order to allow workers to do their jobs carefully and precisely without compromising safety.

◆*Participation at the National Government Level* Unions and affiliates participate in national governmental policy-making to bring workers' rights and needs into public discussion and consideration. For instance, in Germany, such actions helped to improve the regulations for 'fictitious self-employment' jobs, for people with (negligible) part-time employment, and for an adequate 'bad weather compensation' for workers in forestry, in agriculture and on construction sites.

◆*Participation in policy-making at international levels* – Forest workers unions have taken part in the negotiations for different International Labour Organisation (ILO) Conventions (see below), as part of the ILO's tripartite process, as well as in the Intergovernmental Panel/Forum on Forests, in order to promote the recognition and codes of practice on safety and health and basic minimum standards in forestry. Unions are also influencing certification processes in order to have at least requirements of the Core ILO Conventions respected. At local levels of certification or developing certification standards, forest workers are key stakeholders and providers of information in their forest enterprise.

◆*Multinational Framework Agreements* – Under the terms of these agreements partner enterprises demand that suppliers ensure that their workers enjoy conditions which comply with national legislation or national agreements and have unrestricted rights to join trade unions and to engage in free collective bargaining. They demand that their suppliers respect ILO Conventions and Recommendations relating to their operations, such as the 'ILO Code of Practice on Safety and Health in Forest Work', Conventions 29 and 105 on abolition of forced labour, 87 and 98 on the right to organize and negotiate collective agreements, 100 and 111 on equal remuneration and non-discrimination and 138 and 182 on child labour.

◆*Participation of women within the forest sector* Women are also increasingly given attention in the forestry professions, since they have entered at all levels of forest related work, but often remain overlooked and neglected. Their role is quite often different to that of men, and they may have particular interests and issues and different ways of participating in a participatory process. Trade unions work towards the acceptance of women as equal partners in social and economic development.

PROTECTING WOOD WORKERS: THE ILO CONVENTIONS

Within Europe there is a high diversity of forest ownership and forest working conditions. An important basis for worker protection is given through the ILO which is a tripartite body composed of government, employer and worker representatives. The ILO's 180 Conventions and their supporting recommendations provide essential international guidance and protection on many labour issues.

The most fundamental human rights are covered in the seven core ILO Conventions which aim to prevent the very worst forms of repression, exploitation and discrimination. These are amongst the most highly ratified of all ILO Conventions: Conventions 87 and 98 on the rights to freedom of association and to bargain collectively; Conventions 29 and 105 on the abolition of forced labour; Conventions 111 and 100 on the prevention of discrimination in employment and equal pay for work of equal value; and Convention 138 on child labour. Taken together, these core Conventions form a basic minimum level of protection for workers. In addition, the ILO Code of Practice on Safety and Health in Forestry Work which was approved in 1998 provides comprehensive guidance and is now forming the basis for national initiatives in several countries.

CHANGING WORKING CONDITIONS

Globalisation, technical change, erosion of basic rights, 'social dumping'[20] and changes in economic and social welfare form a backdrop for many of the issues affecting forestry workers in Europe. This can be a problem in Eastern Europe and is particularly critical when the newly moved workers do not speak the local language and have no idea of their rights. Apart from the protection of wages and working conditions, other linked concerns of workers and unions include:

◆ *Use of technically sophisticated equipment:* which can cause job losses, and generates demands for retraining; the acquisition of new skills; life long learning and job flexibility. Increased mobility in the workforce can mean a lack of investment in training and increased health and safety problems.

◆ *Health and safety issues:* occupational workplace health, safety and environment committees have been well established in many countries and form a model for providing worker education and training programmes. To supplement regulatory protection of workers many companies have entered into voluntary agreements with unions such as the agreement with IKEA and Faber-Castell and the IFBWW to protect basic worker rights.

◆ *Environmental quality:* for forest workers it is obvious that sustainable jobs are only possible when the forest resource is managed sustainably. Unions are also actively involved in forest certification activities.

◆ *Restitution and privatisation of state-owned assets in eastern Europe:* changes to forest ownership and management in Germany has many implications for workers. In particular, there are moves to decrease the forest workforce, and increase the use of contract labour and piece rates – issues often associated with lack of basic worker protection. Unions are working to ensure that the new private contracts are as good as the older public contracts.

In response to changing work conditions and new worker needs, unions in many countries are providing new services to their members such as help with pension schemes, financial services and discounts on consumer goods. Unions are also merging to form new larger more powerful organisations. For example the forestry and wood unions in Sweden merged in 1997, and the wood processing and metal unions in Germany merged in 1999.

◆Woodworkers are key stakeholders in sustainable forest management, but have often been overlooked in decision making processes from local to international levels.

◆European Unions of woodworkers have developed to promote better working conditions, protect wages, and encourage education and training programmes, etc, at all levels of decision making.

◆The participation of forest workers and unions in forestry decision making is essential for ensuring that the social issues of workers' health, safety and equity are included in forest management.

◆Since forest workers implement forest management decisions they are in a good position to monitor sustainable practices and they should be included in this very important activity, particularly in forest eco-certification processes.

◆Women working in forestry face special issues that are often different to those of men, and these issues need to be addressed as a priority.

SMALL FOREST OWNERS IN EUROPE: THE ROLE OF ASSOCIATIONS IN FOREST MANAGEMENT[21]

INTRODUCTION

About 65% of European forests are privately owned. Some 12 million families from northern Scandinavia to the southernmost parts of Spain and Greece, own and manage forests. The average private forest holding in Europe is 10.7 ha.[22] Around 54% of the raw timber traded in Europe is produced on privately owned forests. Small forest owners in Europe have organized themselves into numerous voluntary forestry associations, which have enabled individuals to overcome the limitations of small and fragmented private forest holdings typical of Europe's forest estate. While the discussion below may not reflect community forestry in its more traditional sense, forest owner and producer associations embody principles of

community involvement in other ways. They were often founded at a community level to serve local forest management and marketing needs; they have institutionalized democratic systems of decision making and representation, and many have become extremely profitable enterprises generating high returns for their members. Such patterns of organization are of interest to countries in transition in central and eastern Europe, many of which are returning forests to individuals.

HISTORICAL DEVELOPMENT

In some countries, such as in southern Sweden, small forest owners – inspired by energetic and visionary community leaders – initiated closer cooperation amongst themselves, and founded local and regional forest associations. State forest administrations have also encouraged private forest owners to establish forest owner organizations. These associations have continued to evolve over time. In the first half of the 20th century most timber produced in small scale private forestry was used for home consumption, such as building, replacing farm buildings and fire wood. In the early 1950s, timber production for local and regional markets gained in importance. While many of the small forest owners associations have their organizational roots in the agricultural sector, such as the farm marketing cooperatives which emerged in the late 1800s, many have now developed specific forestry services, such as the organization of marketing, provision of credit schemes to joint management, and common use of machinery. In the 21st century the number of forest owners living in urban areas is forecast to increase, and the forest owner associations will take on more of a consultative role to this new clientele.

BENEFITS OF COOPERATION

Forest owner associations are based on a belief that cooperation between families can overcome the disadvantages of the small size of individual forest holdings. Forest management depends a great deal on the accessibility of the forest stands. Due to the small sized holdings

PLATE 16: Private Forest Owners in Sweden
(Photo: Skogsägarna)

the burden of investment for a forest road or path would be too great for one single forest owner. In association with neighbouring forest owners the costs of the needed infrastructure can be shared. In most European countries only forest owner associations are eligible for state aid concerning infra-structure investments to improve forest management. When plants or seeds for reforestation or afforestation are needed forest owners associations negotiate for better conditions and prices than the single forest owner could do. The same principle applies to the use of machinery. Furthermore, cooperation between forest owners also increases their competitiveness on the market, and helps them provide the market with the needed amount and quality of timber. The forest owner associations also provide training and information on forestry management and technologies, helping individual forest farmers keep up to date with changing ideas and practices.

ORGANIZATIONAL DIVERSITY

There is a wide variety of forest owner organizations throughout Europe, providing a diverse range of activities and services. National level owner organizations usually have political objectives and represent their forest owner interests in various fora. For example, the Norwegian Forest Owner Federation, with some 57,000 members, represents member interests at a national level, in addition to providing them with technical support. There are also a number of transboundary organizations, such as the Confederation of European Forest Owners (CEPF), which represent owner interests at international policy making levels. At a regional and local level, many forest owner associations have economic, management and information objectives. For example, the forest owner associations in Germany help coordinate timber market activities, provide information and extension services to their members. Some are

103

even involved in the management of the forest estate. In Scandinavia the wood processing units belong to the forest owner associations. However, institutionalized structures at a local level are often lacking, and need to be further developed. Most forest owner associations have voluntary membership systems, and are self, rather than externally organized. Involvement in decision making is usually through a general assembly of members, who elect a board of executives and chairman or manager.

EXAMPLE OF A PRODUCERS' CO-OPERATIVE[23]

A forest owner association may be a producer co-operative. This means that although members own private forest land and make individual decisions about it, they are joint owners of a forest association through their capital contributions. They elect representatives to attend annual general meetings of the association. The representatives appoint a board of directors of the association, which in turn appoints an executive manager. Voting rights vary between associations, but is often based on the principle of 'one member, one vote'. Members share the operating surplus between them, in relation to the volume of timber contributed.

PARTICIPATION IN THE CONTEXT OF SMALL FOREST OWNERS

There are two kinds of public participation related to private forests: participation of the public in or concerning forestland; and participation of the private forest owners in processes organised by themselves. At the forest management unit level, the ability of the public to participate in the owners' decisions is subject to the owners' perceptions of costs and benefits, and limited by their property rights. Many factors mediate the perceptions of costs and benefits, such as the degree of wider social interest or concern; the owner's individual interest in wider issues, such as biodiversity conservation; availability of information and external support; financial incentives and costs; and so on. In general, private forest owners stand to benefit from greater participation in forest decision making. Public participation in private forests may open new sources of

public interest for forestry and offer additional income for private forests owners as well as the possibility to discuss wider and changing public values and means of adapting to new forest practices. It also allows the wider public to better appreciate the investment challenges of private long-term sustainable management. However, in order to be fair, small forest owners believe that participation must be on a voluntary basis, and that clear and agreed ground rules are required for participatory processes.

FOREST CERTIFICATION AND SMALL FOREST OWNERS

Throughout Europe, family forestry is generally considered to be underpinned by the principles of long-term responsibility, tradition, ownership from generation to generation, care for the environment, and democratic accountability within forest associations. Small forest owners are committed to the various resolutions of the pan European Process for the Protection of Forests, which has emphasized the principles and criteria of sustainable forest management (see Part IV). However, spokespersons for small forest owner associations have expressed concerns over the trend towards independent forest certification as a means of validating their sustainable management practices. Many believe that existing forest certification schemes are unfeasible and not adapted to European conditions which are characterized by small scale and fragmented holdings. Some organizations are unhappy about the Forest Stewardship Certification programme, which can appear better adapted to large scale industrial forestry, and may seem to burden small producers with unnecessary costs, bureaucracy, and rules. Because small forest owners are often in a disadvantaged situation, it is feared that this scheme could push European family enterprises out of the market. Where they are less organized at a local level, they tend to lack resources and knowledge about certification procedures. It is felt that this can lead to unfair processes of participation particularly when faced with articulate and well organized environmental NGOs. Since 1999, the small forest owners of Europe have promoted their own pan-European certification process (PEFC), based on resolutions agreed within the pan-European process, recognizing its criteria of sustainable management.

Many environmental groups have, in their turn, rejected this programme, arguing that it certifies the status quo; undermines global solutions by focusing on Europe; and fails to recognize indigenous peoples rights. The debate continues.

ENABLING ENVIRONMENT

The development of the small forest owner associations within Europe has depended upon a number of factors, particularly state support and a developing wood market. Supportive policy and legal frameworks, including flexible and secure legislation underpinning private land ownership rights, the rights to organize and the rights of organizations, are important prerequisites for building such associations. Many small forest owner associations have also depended upon the active involvement of the state, particularly through the activities of the state forest services. For example, some have received financial support, including tax reductions for economic activities, and property tax reductions for members of cooperative bodies. Direct incentives in the form of cost sharing for organizational establishment, staff, investment in equipment or infrastructure and management. The state has also provided organizational support, such as education, training and administrative support for evolving organizations. The development of small forest owner associations has also depended upon a growing demand for wood from the timber industry. The concentration and globalization processes in the wood industry, with the tendency to form even larger units, encourages the development of larger wood marketing units on the supply side, encouraging private forest owners to collaborate.

SÖDRA

Södra is a small forest owner association with 33,000 members in the south of Sweden, who together own 1,700,000 hectares of productive forest land, with an output of more than 10 million cubic metres of timber per annum. Södra consists of a co-operative forest owner association and a wood processing company: Södra Skogsägarna, which is 100% owned by its members. The objective of the Association is to 'endeavor to bring about a secure and meaningful outlet for the entire wood produc-

tion of the members at satisfactory prices'. Its commercial department negotiates timber prices annually on behalf of its members, and sells timber to external buyers. It also delivers timber to Södra's own processing plants. The Association is also increasingly involved with felling members' timber. In 2000 the Association felled more than 66% of the timber on private land. The Association's membership department deals with governance issues, meetings, and election of boards. Since 1995 Södra has developed Green Forest Management Plans, taking into account the forests' biodiversity and environmental values.[24] The overall target is to cover 50% of the total member forest area with green plans by the year 2002. This programme currently employs 80 forest technicians with a conservation training.

The first forest owner association developed in Smaland in 1926, and other regions soon followed the example. The smaller associations joined together into a regional association, which later became Södra. The reason for the early alliances was the depression in the forestry sector due to low demand and low prices in the late 1920s and 1930s. The first economic activities of the forest association were to supply firewood to big customers such as hospitals. In the 1940s the association bought and leased a number of sawmills in order to improve outlets for the members' wood. Its subsequent industrial development has concentrated on pulp industries, saw mills and particle board factories. This was motivated by a timber surplus in southern Sweden and by growing demand for these products. Södra currently owns 6 sawmills which produced about 600,000 cubic metres of sawn timber in 1999.

LESSONS LEARNED

◆The development of small forest owners association depends on a clear delineation of private forest property rights and responsibilities; and government policy and legislation which supports the rights to form interest and technical organizations.

◆Governments have often been influential in supporting the development of small forest owners associations through direct and indirect financial incentives (e.g. subsidies and tax reductions); and providing technical and management support through the state forest agencies.

◆ Associations of small forest owners were often founded by local leaders to serve local community needs in rural areas, and were developed to overcome the diseconomies of scale with regard to supply, production and marketing.

◆ Technological change and involvement in industry have allowed small forest owner associations to develop and adapt to new conditions and to become profitable enterprises.

◆ Membership organizations are voluntary, and have institutionalized democratic systems of governance which are accountable to their members. The wide variety of organizational forms within Europe have evolved in response to local contexts and needs.

◆ Well functioning local and regional forest owners associations, that adapt to the changing demographic and social patterns of forest ownership in the 21st century, can help safeguard and further develop the multifunctional role of forests in support of rural development.

COMMUNE FORESTS IN FRANCE: OPPORTUNITIES AND CONSTRAINTS FOR COMMUNITY INVOLVEMENT[25]

A BRIEF HISTORY OF THE RELATIONSHIP BETWEEN FORESTRY AND COMMUNITY IN FRANCE

The Code Forestier of 1827 stipulated that the State would assume responsibility for the management of most communal forests. At this time, over two-thirds of the population of France lived from agriculture and 70% to 80% of that population lived from less than a hectare. The forest was a direct source of food, fiber, fertilizer, pasture, fuelwood, hunting and many other forest products, and was especially important to small land-owners. The forest became progressively a national good and local claims related to agricultural and pastoral uses were largely curtailed. Insurrections against foresters were common, and afforested land was often deliberately burned. Mountain communities have suffered from the divorce between forestry,

agriculture and pastoral livelihoods in particular. Resistance to subjecting forests to the Forest Regime was common among municipalities during the 19th century.

In the 20th century conflicts receded because of widespread rural-urban migration resulting from the industrialization, the intensification of agriculture and declining dependence on forests. While this process induced an increase in forest cover, forests became marginal in terms of national revenue. Forestry became a mere attachment to agriculture administrations and foresters felt they needed their own autonomous services. In 1964 the *Office National des Forêts* (ONF) was established to manage all public forests submitted to the Forest Regime. Even though the decentralization law of 1983 devolved much decision-making power to municipalities concerning land-use planning, the Forest Code of 1985 still maintains that: "the politics to enhance the economic, ecological and social values of the forest is the competency of the State".

COMMUNE FORESTS TODAY

Forests currently cover about 30% of French territory. Overall public ownership amounts to about 30% of the total forested area, the rest being private landholdings. A total of 11,000 communes own 57% of the country's overall public forested land or about 2.6 million hectares. In the eastern regions of the country and in mountainous areas in particular, communal forests often constitute most of the public forest. In 1995, over half the annual earnings from public forests came from forests owned by communes. After the storm of *Lothar* (December 1999), all French forest communes have suffered substantially from lost revenues, as they all stopped wood extraction in an act of solidarity, in order not to oversupply the market.

The Federation of French Forest Communes was established in 1933 and is the main instrument for municipalities in their continuous demand to be fully considered as owners responsible for the management of their forests and to be better represented in national and regional forest policy-making processes. The

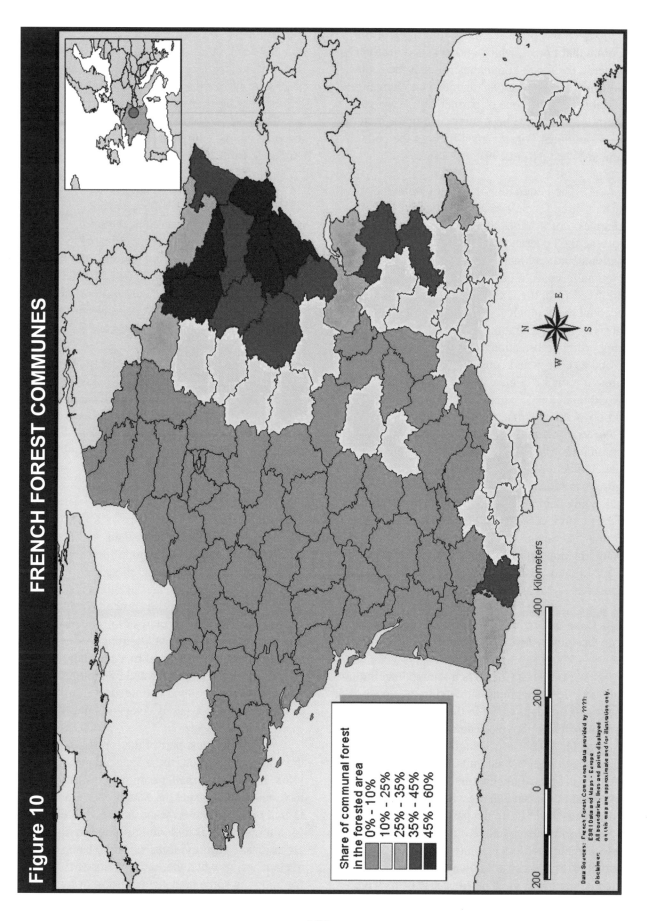

Figure 10

FRENCH FOREST COMMUNES

Share of communal forest
in the forested area
0% - 10%
10% - 25%
25% - 35%
35% - 45%
45% - 60%

200   0   200   400 Kilometers

Data Sources: French Forest Communes data provided by ????;
                ESRI Data and Maps - Europe
Disclaimer: All boundaries, lines and points displayed
                on this map are approximate and for illustration only.

107

Federation has produced the White Book of the Communal Forest, which outlines its objectives. In general, the forest communes ask that their relationship with the national agency is no longer one of submission but one of partnership. Many of their demands have been integrated in a Charter of the Communal Forest signed by both the Federation and the ONF in 1991.

The Federation of Forest Communes is still pushing for precise and progressive policy changes. For instance, in matters of mountain forests, the Federation asks for representation in national and regional commissions working on mountain issues, requesting that forest mountains are freed from the tax burden (25%) taking into account the ecological and social services they provide. The Federation has been innovative in supporting certification, eco-taxes, the use of wood as energy and construction material, measures to mitigate climate change and in promoting wood related professions. The Federation is proposing a new management concept called *Chartes de Territoires Forestiers*, for the communes to integrate forestry into larger rural development objectives, and to involve interest groups in sharing the costs and benefits for achieving commonly agreed objectives. The Federation is also actively involved in providing training to forest commune officials. It is a founding and active member of the European Federation of Commune Forests and the European Observatory of Mountain Forest.[27]

THE CASE OF SIXT: CONSTRAINTS AND OPPORTUNITIES FOR COMMUNITY INVOLVEMENT

Sixt Fer à Cheval is a mountain commune of about 750 inhabitants situated in the upper valley of the river *Giffre* in the French Alpine region of Haute-Savoie. The commune has a territory of 11,700 hectares, ranging from 770 to 3099 meters at its highest summits. The incredible richness of the fauna and hunting possibilities attracted nobles and clerks whom built a Monastery in 1144. In the past the forest was key to local subsistence and livelihoods, and resource use was more or less regulated, usually by the municipality. Sixt was known to be a relatively rich commune in great part because of its large

and productive forests.[28]

The commune has had most of its territory (9,200 ha.) under the status of nature reserve since 1977, with objectives to protect species and habitats. The reserve is not inhabited – except by some seasonal guards and visitors of refuges and herdsmen for seasonal pasturing.

In 1993 the commune has been also selected as *Grand Site National* thanks to its cultural, aesthetic and ecological qualities. The commune is proactively using the label to obtain financial contributions from national and regional governmental agencies, for developing the Monastery into a museum and larger information center on the Nature Reserve, and to enhance the value of other sites in the commune.

The village has still relatively few hotels and secondary residencies and few skiing installations. Even though it attracts some 300 thousand tourists a year, Sixt does not greatly benefit from this summer, short-term affluence. Many think that the development of skiing facilities would provide a great asset.

THE FOREST TODAY

The commune owns about 3,000 hectares of forests. Most of the commune's territory, except the narrow valley bottom and mountains over an altitude of 2,000 metres, are forested mostly with beech *(Fagus silvatica)* and spruce *(Picea abies)*. There are about 1800 hectares of private forests, much of it being very small holdings, often without clear boundaries and owned by entire families. Nearly all public land is owned by the municipality. State-owned land is limited to some risk prone areas, such as along rivers (about 300 ha.).

The greatest part of the forest belonging to the commune is under the management of ONF (1800 ha.). Some smaller commune forest patches have remained outside the State Forest Regime. Usually the communes want to reserve such forests for alternative uses including: pasture in higher elevations; protection forests against land slides or avalanches; or when the forest is close to villages for allowing some housing extension or tourism

PLATE 17: The mountain commune of Sixt, France.
(Photo: J-J. Richard-Pomet)

development. The protection against avalanches and landslides is a key preoccupation of locals and a cause of conflicting views between them and ONF. Local people have tended to leave the forest untouched in risk prone areas, while ONF considers that the aging forest should be rejuvenated to better fulfill its protective function.

High production costs, especially in mountains with difficult terrain, and ever declining wood prices provide few incentives for local communities to get involved in forest management. Compared to other European countries the French State gives between four to ten times less subsidies per hectare of forest.[29] In this context, ONF as well as other forest owners have little incentive to value the environmental and social benefits of forests as much as their wood production function.

ONF is at the end of a forest management plan that was established in 1983. Even though the plan has to be signed by the mayor to become

valid, it has obviously been entirely elaborated by ONF. Indeed it is written in authoritative terms towards the municipality. It stipulates that the commune should relinquish 50% of its forest revenue, identifying insufficient investments in forestry by the commune during preceding years. However, in 2000, the commune still invested in its forests (FF 110.000) even though it obtained no substantial income from it because of the fact that logging activities were frozen following *Lothar*. The commune should receive a credit from the State in order to compensate for part of this loss.

There are signs that the technocratic approach to forest management of ONF may be changing. A new generation of foresters seems to value more multifunctional forestry and participatory approaches. A young local forester in Sixt described his job as "managing light in the forest and communication".[30]

The local people are still interested in fire-

109

wood. It is the tradition of French forest commune to offer its residents access to woodlots for cutting firewood. This system called *affouage* was formerly free but for a few years the commune has decided to charge 250 FF for 6 steres of uncut timber. One of the given reasons is that when *affouage* rights were entirely free the tendency was for forest owners to prefer to exploit communal woodlots, rather than using – thus "cleaning" – their own forests. Furthermore, while wood from *affouage* should traditionally only be used domestically, the commune could not prevent people from selling the fuelwood at a substantial profit. Firewood still represents an important source of income for some local people.

There are currently three sawmills in the valley, four chalet builders and three carpenters. However, the local wood industry uses very little local wood. The spruce exploited by ONF often goes to further destinations and hardwood timber remains largely unused.

COMMUNITY BASED LAND USE MANAGEMENT IN SIXT

While wider social and economic changes have resulted in the diminishing importance of commune forests for rural development in Sixt, the local commune has been involved in establishing multiple-stakeholder based land use management schemes in which forests play a part.

The Nature Reserve was established by the NGO Friends of the Reserve in 1977 after a public inquiry (required prior to the creation of all nature reserves). The active members of the Friends of the Reserve have been largely elected councillors. The commune actually provides FF 20'000 (US\$ 2632) a year to the Reserve besides offering the infrastructure and some maintenance. Since then the management of the Reserve has been transferred to a broader Consultative Management Committee, which includes the commune, the private owners, environmental protection organizations and various representatives of official administrations, as well as scientific advisers.[31] With a reduced managerial role, the Friends of the Reserve are now mainly involved in educational activities, offering activities for visitors, and discovery classes for the village's school.

It also revives the local cultural patrimony by producing booklets on the history of the Monastery, iron mining in past centuries, etc.

Forestry practices in the Reserve are left more or less unrestrained, other than the ruling that there should not be any clear-cuts over one hectare without authorization of the prefect. The relationship between ONF and the Reserve has often been difficult. Even though full-time residents of the commune are allowed to harvest wood for domestic uses (except for species bound restrictions valid in all territories) and to hunt (without dogs) on about half the reserve's territory, some landowners and hunters still perceive the Reserve as a constraint. Some residents feel they should be freer in restoring the huts in mountain pastures. Local people actually show little motivation for being involved in the management of the reserve.

THE CONVENTION FOR THE RESTORATION OF PASTURES AND MOUNTAIN DAIRY ECONOMY

One of the major issues in mountain communes is the disappearance of agriculture. In Sixt there are only 3 farms left. Compared to Switzerland, France provides little extra support to farmers of mountain areas. The commune is concerned about the disappearance of farms given their role in shaping the valley landscape. Fourteen partners have developed a Convention to restore mountain pasture, and its cheese dairy within the Reserve. The partners include the commune, the keepers of the pasture, the Friends of the Nature Reserve and the reserve's scientific committee, local and regional hunting associations and some officials (agriculture and forestry administration). The commune also received a contribution from Natura 2000 to help rehabilitate the dairy farm. The Convention's objectives are economic, social and ecological. The decision-making power is shared among all parties to the agreement while the commune is the manager of the project.[32]

ASSOCIATION OF PASTURE OWNERS

These days the pasture owners are usually forest owners, since pastures have been overgrown.

The main objective for establishing an *Association Foncière Pastorale* is to integrate the management of a very divided private property domain, in particular to clear some forests or overgrowth of vegetation from former pastures, for landscape and protection purposes. The commune is proactively involving the local owners by contributing itself over half of the land (3500 ha) that should be managed by the group.[33] It may be surprising that the commune plays here the role of a private forest owner, but by law, communal forests are the private property of the commune.

LESSONS LEARNED

◆The legacy of feudalism, the Revolution, the philosophy of Enlightenment shapes French culture with a strong edge for State-centered and technocratic administration. However, this same legacy also provides a solid corpus of laws, which, if well used, can provide the public with powerful means to defend its interests. For instance, according to the latest evolution in the general Code of Territorial Collectivities (1996), residents can participate in their commune by being associated with public projects, through public inquiries, referendums or the creation of consultative committees. However, in the case of Sixt, these means of direct public or community participation are only timidly used, while partnership agreements involving community representatives with some organized stakeholders are more actively promoted.

◆The case of Sixt indicates that, for French mountain communes, forests tend to be less a key asset for rural development than they have been in the past. Although forest cover is increasing, use is declining and, given the current institutional context, of relatively little interest to locals. The mountainous terrain, low timber prices and high production costs further constrain CIFM. The situation has become even more dramatic for the French timber market since the storm *Lothar*. Safeguarding agriculture and pastoral activities, valuing the local ecological and cultural patrimony while promoting tourism are currently more important to the commune.

◆Municipalities are also governmental bodies.

Once elected they tend to turn their attention towards other governmental agencies upon which they depend financially, sometimes neglecting to foster opportunities emanating from local people and resources. However, the proactive role of the Federation of Forest Communes and the case of Sixt, show the commitment of local municipalities to enhance their role as main decision-makers in local land and forest resources management.

◆There is also a tendency for citizens in modern society to be less concerned by local affairs, because they are less dependent on local resources for their livelihoods, and because they develop their cultural identities and social networks beyond local areas. However, the case study of Sixt suggests that there is a renewed interest of its young people in staying and working in the commune.

◆The overlap of institutional claims over the same territory creates governance problems. This is problematic for the concerned organizations and their relationships, and also complicates the participation of the public in the management of that territory. Conflicts over the uses of a territory and its resources also occur at higher administrative levels, the commune being caught in between their diverging claims.

◆There is a renewed interest in integrating land use management on a territorial basis as illustrated by the trend towards multiple-stakeholder natural resource management systems, such as Regional Natural Parks, Charters for Forest Territories and other conventions. In fact, linkages between rural and tourism development are often key in these institutional innovations. The challenge is to value forests more and to involve communities within these schemes more directly.

ENCOURAGING INVOLVEMENT IN PUBLIC FOREST MANAGEMENT: EXAMPLES FROM FINLAND, DENMARK AND SWITZERLAND

FINLAND: PUBLIC PARTICIPATION IN FOREST AND LAND USE PLANNING[34]

Metsähallitus, the Finnish Forest and Park

service (FPS) made the decision to use public participation in forest planning processes[35] about five years ago. The goals of its regional forest and land-use planning in the mid 1990s were to:

◆ Generate a widely acceptable land-use plan, allowing national goals to be integrated and harmonised with the aspirations of the general public.

◆ Gain information on various stakeholders and develop good working relations with them.

◆ Encourage individuals and interest groups to participate in the planning process.

◆ Learn collaboratively about the goals and objectives of all stakeholders towards the use of state forest.

◆ Understand major issues and concerns related to natural resources and their management in the region.

◆ Inform the public about the FPS and the services and opportunities provided by the agency.

◆ Utilise local knowledge.

◆ Integrate public participation into the agency's everyday business.

The process was initially motivated by a desire to address conflicts occurring on FPS land. At the time there were no laws obliging the FPS to undertake a participatory planning process; although since 1997 the new Forest Law advocates the use of public participation to serve nation wide planning processes.[36] The changing legislation reflects an institutional appreciation of the principles of open and cooperative ways of working.

The participatory process was designed by a few FPS participation specialists, but was adapted in response to evaluations and proposals throughout the process. The main methods for involving the public included:

◆ Media, news announcements and brochures on the process.

◆ Open houses (x4).

◆ Information access points at customer service offices (x6).

◆ Public meetings (x12).

◆ Opportunities for comment via feedback forms, letters and paid phones.

◆ Personal contact between employees and individuals.

Members from the wider public and organised interest groups were involved in the process from the beginning, including representatives from communities, cities, forest and environmental centres and associations, the army, citizen groups from villages, anglers, hunters, ornithologists, local businesses, workers, schools, universities, reindeer owners, etc. Some 400 interest groups were invited to the process and about 150 participated in the first meetings when working groups were established. The main goal of these groups was to represent different points of view and geographical areas. The groups provided fora to negotiate and build consensus for the land use plans. Working groups met for all the important decision making points in the planning process. There were about 6-8 meetings per group. Most group representatives were men (75%) of middle age, and of higher education.

A preliminary evaluation indicated that the public participatory process has provided fruitful ways of collaborating and co-operating with the wider public and organised interest groups. It has enhanced an understanding of the diverse opinions and goals of many stakeholders. The new planning approach has allowed the public to overcome their image of a closed government bureaucracy, and made people feel more comfortable about contacting FPS staff. Although some people have questioned the effectiveness of their participation, research suggests that some 80-90% of the ideas from the public have been integrated into the final forest and land use plan.[37] While Metsähallitus may be the final decision maker, working groups can take decisions about alternatives throughout the process. Some FPS employees expected greater public demand for nature protection in old growth forest, criticism on public forest management practices, and fear that increased media attention would further degrade the credibility of the agency. However, nature protection generated less public input than was originally anticipated. People focussed more on living conditions, employment opportunities, outdoor recreation opportunities, such as fishing, hunting and berry picking. The media were generally supportive of the participatory approaches being adopted by the FPS.

Public participatory processes in Finland do face some challenges. Agency employees recognise that participatory processes involve extra work and financial resources, and a sustained commitment to a long term process. Planners sometimes get frustrated because of lack of public interest in forest planning. Long processes also experience difficulties in keeping all groups interested all the time. The use of specialist jargon is also a problem. Furthermore there may be a backlash to public participatory processes from the influence of the global economy, generating some feelings that Metsähallitus should not be so open. One explanation is that some groups use the open management strategy to collect information for private use in the media. Despite these reservations, research indicates that FPS employees are convinced that participatory approaches are the best way of working in the future

Forest Agency Changes. Several FPS employees received education in participatory processes and conflict management in the USA prior to undertaking the regional planning exercise. They held training workshops in Metsähallitus for other staff members dealing with the role of public participation in planning. During the regional planning process, 3-4 specialists worked on the participatory planning full time. Although FPS aims to get local employees familiar with participatory processes as fast as possible, it has been advantageous to have one or two specialists working with planning and education full time. Public participation has not involved an increase in the permanent employees in FPS but relied more on visiting specialists to transform skills within the agency. There has also been need for larger education programme on "people skills". It

has been important to find people who feel comfortable in these situations.

Metsähallitus' working climate was very closed in the beginning of 1990s, but since 1995 it has experienced big changes, with the organisation developing its own more participatory culture. In fact, open co-operation has also been written into Metsähallitus' own common values. All internal planning processes (strategic land use planning, landscape ecological planning, zoning for state owned shores, etc) have radically changed as a result of public participation. Participation has generated working groups representing many interests, public meetings, open houses and so on. Metsähallitus has developed many new partnership as a result of public participatory process. The planning timeframe has also become much longer because of participatory processes. Metsähallitus' own planning teams have changed to include groups working dealing with participatory processes; timetables; meetings; feedback collection and analyses; research; public education etc. The success of public participatory processes requires a positive attitude amongst agency staff and new skills for interacting and working with the public.

DENMARK: USER COUNCILS WITHIN STATE FOREST DISTRICTS[38]

In 1995 the Danish Forest and Nature Agency introduced 'user councils' within each of its 25 state forest districts. The goal of the councils is to enhance the involvement and influence of local users – the 'common citizen' – in the management and utilisation of public forests. They were established to allow participation in the planning of state forests, which takes place every 15 years. Until the mid 1990s public participation in State forest planning had been very limited. Forest policy and administration had been expert-led with permanent or ad-hoc advisory boards found at the national level. Forest planning had included hearings of major NGOs and concerned municipalities and counties, but had not included wider public involvement.

In 1995 some 33 user councils were established with up to 14 members in each. In practice the user councils have ended up as a mixture of environmental NGOs, country officials, municipal politicians and – in some councils – representatives from defence, agricultural, hunting and tourist organisations. The majority of members are male (83%). The average age of the council members is 54. Since 1998 the user councils have met at least twice per year. They have no formal decision making power, and it is up to the forest supervisor whether s/he involves the user council before or after a decision has been made. The councils are widely used for informing and consulting with the public on forest management decisions, although the user councils could potentially develop partnerships with the state forest agency.

In general it is felt that the user councils have facilitated communication and increased understanding of actions and motivations among different stakeholders and state forest districts. Further, the improved dialogue with municipalities and farmers' organisations provides opportunities for afforestation and nature restoration projects in line with the Danish ambition to double the country's forest area. However, the success of the user councils depends on whether its members feel they can make a difference. This relies on the forest supervisor facilitating communication, transparency and accountability. The major limits of the user councils could be seen as the forest supervisor being unwilling to manage the user council; that the current composition of members fail to represent the forest users, and lack of interest among citizens participating.

SWITZERLAND: REGIONAL FOREST PLANNING IN THE LAKE DISTRICT, CANTON OF FRIBOURG[39]

Under the new Swiss Law on Forests, which took effect on 1 January 1993, public participation procedures were formally introduced in forest management planning activities at regional level.[40] These Regional Forest Plans (RFP) are of strategic use in management planning, setting medium and long-term objectives that address society's interests in forests beyond the scope of a single – public or private – forest estate. They represent a framework for future orientations and decisions of forest authorities. In Switzerland, the cantons are responsible for enforcing federal laws within their territory (subsidiarity principle). This implies that the methods of public involvement in regional forest management planning vary considerably from one canton to another.

The canton of Fribourg was one of the first to revise its regional forest plans under the new forest law. The overall goals of the participatory process were to incorporate different public interests and help create a sense of joint responsibility for the forests; to build consensus between different actors; to provide information on complex forest systems, and to create a forest lobby among the population. As a specific management instrument the objectives included increasing timber harvesting and protection of natural elements; finding solutions to wildlife problems; improving infrastructure and recreational aspects; involving schools, and enhancing relations with private forest owners.

The district forest engineer and the project engineer (both women) initiated the public participatory process, and were aided by an external facilitator. Representatives of many different organisations were invited to participate: leisure and recreational associations; conservation and nature protection NGOs; officials, politicians, municipalities, private forest owners, and so on. At a first meeting the duties and special skills of all actors were distinguished, and expectations of participation were clarified. Many different participatory methods were employed during the project over a 17 month period, including role play. The results were integrated in a 14 year regional forest management (1996-2010) in a transparent way, and presented to the participants as well as the public. The media followed the whole process. It is felt that the public participatory process enhanced mutual learning and respect between stakeholders, and has formed a new network of interested groups. Its success was based largely on the openess of the forestry professionals to new ideas, and the assistance of a professional moderator.

Changes within public forest agencies. The recent introduction of public participatory procedures in RFP and the heterogeneous application of this concept in the various cantons, make it difficult to evaluate actual changes which have occurred to date. However, based on first experiences, one can consider following elements:

Regional forest management planning has introduced *new forms of planning procedures and decision-making strategies*, based on dynamic participatory models and modern communication techniques (e.g. working groups, workshops, public meeting, open houses, broad media coverage, etc.). This *operational transformation* of forest management planning has stimulated public involvement in forest decision-making and has improved social acceptance of regional forest management outcomes, thus also improving their quality. However it has also increased the planning time required, the costs and the complexity of RFP. This change also sets new requirements in participatory management (i.e. negotiation and mediation) and places higher demands on the communication capacities of public forest agencies. Even though regional forest management planning has not induced major *changes in the organisational structure* of public forest agencies (e.g. no additional staff recorded since then), they have improved cooperation with neighbouring sectoral public policies concerned by RFP procedures (e.g. nature conservation, land management, agriculture, etc.). It has also increased collaboration with other fields of expertise concerned by the question of public involvement in decision-making processes (e.g. sociologists, policy scientists, mediation and communication specialists). Further, even though final decisions remain in the hand of forest authorities, regional forest management planning has induced a *functional transformation* of the mode of intervention of forest services. They now increasingly seek to build consensus, exchange information and to reach social acceptance in forest decision making.

LESSONS LEARNED

◆Public participatory processes have been increasingly adopted by state forest agencies across Europe for forest planning purposes since the mid

1990s, allowing diverse stakeholders' perceptions and needs to be integrated into forest plans.

◆On the positive side, such processes have clearly enhanced mutual understanding, and improved dialogue between the various stakeholders. They have helped identify local community needs, and generated many ideas for forest planning. However, further research is required to evaluate whether these processes adequately identify and integrate community aspirations and needs.

◆There are many challenges to public participatory processes. Experience indicates that the composition of user councils and working groups usually consist of organised groups, and may fail to represent communities, or a full range of forest users. They tend to be dominated by older men. State forest agencies continue to have the final word about forest planning, and it remains to be seen whether the local groups can develop real, lasting partnerships with government forest bodies.

◆Increasing interaction between society and the forest sector has a broad influence on public forest agencies' institutional roles and functions. Public forest authorities are increasingly called on to play a major role in integrating and balancing the various societal demands on the use of forest resources and spaces. It can be very time consuming and costly for public agencies to undertake such projects. It is often difficult to generate and sustain public interest in forest issues. While many forest agency employees are clearly committed to public participatory processes, future success will depend on deeper cultural and organisational changes within public institutions. They will need to increasingly distinguish their emerging mediating responsibilities from their more traditional forest management tasks. They will also be challenged to develop new strategies and mechanisms enabling a long term social commitment to sustainable forest management in a society that is increasingly short-term oriented.

Notes

[1] This case study is based on research by the editor in 1999, but also draws on Merlo, M. (1989): "The Experience of the Village Communities in the North-Eastern Italian Alps" in Merlo, M; Morandini, R; Gabbrielli, A; & Novaco, I. (1989): *Collective Forest Tenure and Rural Development in Italy*. Rome: FAO. Merlo, M (1995): "Common Property Forest Management in Northern Italy. A historical and socio-economic profile", in *UNASYLVA* 46 (180):58-63; Morandini, R. (1989): "Rural Communities and Forestry in Italy", in Merlo, M. et al (1989): op.cit. Thanks also to Bruno Crosignani for his inputs into and comments on this case study.

[2] Thanks to Roland Brouwer for this case study.

[3] Thanks to Lars Carlsson for this case study.

[4] Thanks to Mandy Haggith and Bill Ritchie for this case study. Their references include:

- Brown, K. Crofter Forestry – a report to the Nature Conservancy Council for Scotland. NCCS. Inverness. 1991.
- House of Commons, Agriculture Committee. Land Use and Forestry. Minutes of Evidence. Wednesday 23 November 1988. HMSO, London. 1988.
- The Crofters Commission. A Guide to Crofter Forestry. 1992.
- UK Government. The Crofters.

[5] Thanks to Olof Johansson and Nanna Borchert for this case study.

[6] Thanks to Willie McGhee for inputs into this case study, which also draws on Jeanrenaud, S. & Jeanrenaud, J-P. (1996/7): *Thinking Politically about Community Forestry and Biodiversity: Insider-driven initiatives in Scotland*. Rural Development Network Paper No.20c. Winter 1996/7. London: Overseas Development Institute.

[7] *The Borders Forest Trust 1998 Annual Review*.

[8] Other BFT projects include: The Southern Uplands Partnership; Carrifan Wildwood; Ettrick Habitat Restoration; School Grounds; Conservation of Juniperus communiy & Sciurus vulgaris.

[9] pers. comm. 1996.

[10] Cited in Sutherland, G. (1993:79): *Explorations in Wood: The Furniture and Sculpture of Tim Stead*. No Butts Publishing, Galashiels, Scotland.

[11] Sutherland: ibid:77.

[12] Thanks to Eoin Cox, from Woodschool Limited, for this section.

[13] Thanks to Yorgos Moussouris and Pedro Regato for this case study, adapted from: WWF Mediterranean Programme: *Forest Harvest: An Overview of Non Timber Forest Products in the Mediterranean Region*. WWF International, Gland, Switzerland.

[14] See glossary for definition.

[15] See Part III for a description of Mediterranean forest types.

[16] The compilation of information in the following table is based on data from Algeria, France, Greece, Italy, Morocco, Portugal, Spain and Tunisia.

[17] Thanks to Simon Levy for this case study.

[18] Thanks to Jill Bowling for this case study.

[19] These also include wood workers in their membership.

[20] Social dumping involves moving workers from an area with lower standards to another area, resulting in lowered standards there.

[21] Thanks to Natalie Hufnagl of CEPF for commenting upon this case study.

[22] UN-ECE/FAO (2000): *Forest Resources of Europe, CIS, North America, Australia, Japan and New Zealand*. New York and Geneva: United Nations: p.7.

[23] Example of Södra in Sweden.

[24] Södra (undated): *Green Forest Management Plans*. Södra, Växjö, Sweden.

[25] Thanks to Andrea Finger for this case study.

26 Jean-Claude Monin, Epinal, October 2000.

27 The actual paper has been established in great part thanks to the material and advice generously offered by the Director of the European Observatory of Mountain Forest, Pier Carlo Zingari and by Chrystelle Bertholet, St Jean d'Arvey.

28 Interview with Mme Anna Mogenier, Sixt, October 2000.

29 Jean Louis Bianco – La Forêt: Une Chance pour la France, 1998, p.1.

30 M. Riggi, Agent Forestier, ONF, interview, Sixt, October 2000.

31 Création de la Réserve Naturelle de Sixt-Passy Décret No 1228-77. Mme Caroline Scuri, animatrice pour les Amis de la Réserve et M. Richard Pomet, garde de la Réserve, interview October 2000, Sixt.

32 Convention de Sixt Fer-à Cheval – Réserve naturelle de Sixt Passy, Convention dans le but de pérenniser les alpages laitiers de la réserve naturelle de Sixt Passy et d'augmenter la valeur du patrimoine naturel, Sixt Fer à Cheval, 15 octobre 1997.

33 M. Pierre Moccand, Mayor of Sixt, interview October 2000.

34 Thanks to Pauli Wallenius for this case study, partly adapted from ILO (2000): *Public Participation in Forestry in Europe and North America.* Geneva: International Labour Office.

35 From strategic land use planning to stand level planning.

36 The new Finnish laws for Community Administration and Community Land Use and Construction also promotes public participatory planning.

37 Editor's note: some environmental groups are skeptical of the public participatory planning in Finland, seeing it as a cosmetic exercise, particularly where logging has been undertaken in sensitive wilderness areas in Malativa. For example, see Kunnas, J. (1999): "Finnish Wilderness Destruction", in Grant, K. (Ed) (1999): *Europe's Forest: A Campaign Guide.* Amsterdam: A SEED Europe.

38 Thanks to Tove Boon for this case study, adapted from ILO (2000): op.cit.

39 Thanks to Yves Kazemi for this case study, partly adapted from ILO (2000): op. cit.

40 According to Article 18, paragraph 3 of the Federal Ordinance on Forests (1992): "When planning goes beyond the scope of a single enterprise, the cantons shall ensure that the public (a) is informed about the objectives and the course of the planning process; (b) can be associated in an adequate way; (c) has access to the information".

Part VI

OPPORTUNITIES AND CHALLENGES FOR CIFM IN THE 21ST CENTURY

This chapter reviews the issues raised in Part I in the light of the case studies, and discusses some of the main economic, social, ecological and policy opportunities and challenges facing CIFM in Europe today. It then outlines the principal lessons learned according to three main groups of actors: governments, NGOs, and local communities.

MACRO-ECONOMIC CONTEXT

From an EU perspective, forestry is seen as an important component of sustainable and integrated rural development, although there remain considerable uncertainties related to the measurement of its contribution to rural development – which is itself an elusive concept.[1] The current economic viability of CIFM in Europe has to be viewed in the light of Europe's prevailing macro-economic context, and regional differences. In general, the EU's CAP continues to have a significant impact upon land use choices throughout the EU 15, and agricultural subsidies typically undermine choices in favour of forestry in many countries. With the exception of some commercially successful community forestry enterprises, such as those in the *Val di Fiemme* in Italy, the *Swedish Forest Commons* and *Small Forest Owner Associations*, the long rotation periods in European forestry, and the low timber prices currently make high volume, low value forestry an unattractive land use choice for individuals and communities alike.

LEGACY OF INDUSTRIAL FORESTRY PARADIGM

Within the public forest sector, subsidies, grants, fiscal incentives, extension services, forest management prescriptions and decision making have generally favoured the expansion of an industrial forestry pattern. This has often been at the expense of local communities and natural ecosystems, as examples in Part II illustrated. Under these conditions, timber is often extracted using migrant labour, and with little value-added at source or benefits accruing to the local community. This has contributed to the decline of rural economies with a reduction in local employment, housing and services in many areas. However, loss of community involvement has been offset by a new dependency on employment in industries and the rise of the modern welfare state which has provided material benefits in exchange for loss of local control. To encourage greater CIFM, it is recommended that new policies should promote the reinvestment of revenues in rural communities and their environments, and the adding of economic values to forest products near their point of extraction. Particular attention should be given to marginalised rural areas in Europe.

FINANCIAL INCENTIVES

The availability of financial incentives for community forestry initiatives in Europe is variable. Many small scale community initiatives, such as the *Crofters* and *Borders Forest Trust* in Scotland, rely on start-up grants and donations from a wide range of bodies, highlighting the importance of direct material support for the development of community based projects. However, in order to achieve long term economic viability, many community initiatives are beginning to recognise that they must reduce dependence on external grants and increase the proportion of their income derived from membership, fundraising, and from marketing their own goods and services. Incentives such as tax reductions for economic activities, and property tax reductions for members of cooperative bodies, have helped the development of *Small Forest Owner Associations.*

Many grant schemes have weaknesses. For example, some community woodland grants in the UK are offered for amenity and recreational benefits only, and do not support the livelihood or employment opportunities which could be generated from community woodlands. There is also concern that woodland grants may be benefiting the richer landowners – such as trusts, industrial interests, and large estates – at the expense of smaller farmers. Grants are usually paid after the completion of work, and are staggered over several years, many small farmers without operating capital experience cash flow problems at the start of their programmes, which may inhibit uptake by smaller initiatives. Applications procedures and accounting specifications for some funds are often viewed as an administrative burden, and application for financial support may be outside the capabilities of smaller community projects.[2] The role of subsidies, grants, and fiscal incentives in supporting community forest management projects and local organisations needs to be improved and more widely promoted. Alternative community based models of economic development that embody a full range of values (social, economic and ecological) should be supported, and seen as alternatives to existing centralised and consumptive, economic activity.

THE POTENTIAL OF NTFPS

Despite the broad macro-economic constraints and incentive problems, evidence from the case studies suggests that there is enormous potential for developing new wood and NTFP

processing enterprises within Europe. In many areas forests are a largely under utilised resource, and the development of smaller scale processing technologies, such as the 'tree station' developed by the *BioRegional Development Group* in urban UK, in conjunction with the development of markets and trade, could increase employment opportunities and enhance rural and urban livelihoods. The development of *NTFPs in the Boreal* and *Mediterranean* regions is also known to benefit from the availability of small to medium sized processing equipment.

ENHANCING THE ECONOMIC VIABILITY OF CIFM

The case studies suggest that rural and urban economies can be revitalised through:

◆ increased utilisation of forest and NTFPs

◆ diversifying woodland activities

◆ developing entrepreneurial skills

◆ value-added processing near forest resources

◆ improving infrastructure networks

◆ providing access to capital.

TENURE AND POWER ISSUES

Forest ownership, access and tenurial rights have a powerful influence over the way forests are – or fail to be – managed. They structure the rights of individuals, local groups and wider civil society to benefit from forest products and services. Security of tenure and usufruct rights, provide individuals and rural and urban groups with economic and legal incentives for long-term sustainable forest management.[3] It should also be remembered that tenure defines more than relations between people and property – it also defines social relations between people. Those with tenurial rights have a certain social status vis-à-vis natural resources in comparison to those without rights to those resources.[4] Power relations are deeply rooted in these ownership and tenure systems, and can have an informal but powerful influence on CIFM in the region.

Changing land tenure patterns has had a number of significant implications for CIFM in Europe, and raises the question of whether some land tenure regimes appear more appropriate for CIFM than others – explored below.

EROSION OF COMMON FOREST OWNERSHIP AND RIGHTS

In general, Europe has benefited from a clear delineation of private and public forest property rights and responsibilities.[5] However, as a result of nationalisation and privatisation, traditional common forest properties and rights have been widely eroded since the Middle Ages in most European countries, as land has been expropriated by the State or privatised, causing the marginalisation of many rural peoples, as the examples in Part II , and the case study of the *baldios* in Portugal indicated. Several traditional common forest properties have survived into the 21st Century, and have successfully adapted to new contexts, by adopting progressive commercial and environmental policies and actions, such as the *Val di Fiemme* in Italy, and the *Swedish Forest Commons*. As an alternative to public and individual private ownership, communal forest properties can provide a socially equitable and environmentally sound basis for forest management.

PRIVATE FOREST OWNERSHIP AND CIFM

The high proportion of forest in private ownership in Western Europe provides a special context for CIFM. In some areas, such as Scotland, privatisation has resulted in a very unequal distribution of land holdings, where a small percentage of landowners, own most of the land. It is recognised that land tenure reform is critical in such areas to achieve a more equitable distribution of land rights and to secure wider access to forests. However, experience to date suggests that a redistribution of land rights will be a long term process, and community forest management agreements with the State (concordats) are necessary interim solutions to ensure that forest benefits accrue to local people. Privatisation also challenges the traditional forest usufruct rights of indigenous peoples, such as the *Saami in Sweden*, who are in danger of losing their customary rights

without greater internal and external political support. Indigenous peoples' rights, including the rights and responsibilities to use and manage lands, territories and forest ecosystems within ancestral domains, must be recognised, respected and guaranteed.

Notwithstanding such important exceptions above, compared to many other regions privatisation has resulted in a relatively equitable distribution of land ownership in Europe, and has not precluded other opportunities for collaboration in forest management. Secure private forest ownership has provided the basis for the formation of many economically prosperous *Small Forest Owner Associations*. Their experiences are of potential interest to countries in central and eastern Europe, which have undertaken programmes of forest privatisation and land restitution, and where rural peoples are more sceptical of institutionalised communal forest management. As discussed in the national contexts in Part IV, and the case study on *Small Forest Owners*, attitudes towards public access to private forests, and the role of public participation in private forest management vary throughout Europe, and are intrinsically more problematic because they can appear to challenge private ownership rights. Private forest owners may find it more difficult to incorporate wider sustainability goals of social equity, including supporting indigenous peoples' rights.

PUBLIC FOREST OWNERSHIP AND CIFM

Public forest ownership, until recently, has not provided many opportunities for CIFM in Europe. As the study of *French Forest Communes* indicated, decentralised public forest administration does not always provide systems which actively involve or benefit local people. Tensions between local administrations and more centralised authorities appear to be common in many situations, where their various rights and responsibilities are a matter of constant discussion. However, public forest ownership does not preclude CIFM either. Even street trees, owned by municipal councils in urban areas, can provide resources for community involvement in forest management, as the *BioRegional Urban Community Forestry Project* in London demonstrated. In general, the role of public forests are changing in western society, and forest agencies are paying more attention to the recreational and leisure needs of rural and urban communities, and are beginning to institutionalise public participatory processes in forest planning and management.

In short, private and public tenure systems provide both opportunities and constraints for various types of CIFM. Contrary to some popular opinion, there are powerful social and environmental arguments for maintaining and re-establishing common property institutions in many instances. It is recommended that they should be encouraged in Europe as much as in developing countries where traditional systems have been more recently disrupted. The scope of CIFM based on private and public land ownership systems largely depends on how well wider social, environmental and economic values are integrated into different management regimes.

FOREST GOVERNANCE ISSUES

What kind of governance institutions provide opportunities for effective CIFM? Relatively speaking, the liberal democratic context within Western Europe, with its associated rights and responsibilities, has underpinned the development of a wide range of public and private resource management institutions. For example, associations of small forest owners have depended on political freedom to associate freely, express opinions publicly, form and register organisations, open bank accounts, secure private property rights, obtain information, and to participate meaningfully in various fora.

LOCAL FOREST GOVERNANCE

Several of the case studies, such as the *Val di Fiemme* in Italy, the *Swedish Forest Commons,* indicate that local forest governance can be an effective way of developing socially acceptable rules about access and rights to resources, responding to local needs, aesthetic and spiritual values, knowledges, resolving conflicts, and distributing costs and benefits equitably. In some instances, adaptations of traditional community institutions seem to provide acceptable forms of

local governance for new woodland schemes, such as the use of traditional grazing committees in some Crofting areas in Scotland. Local groups often have effective local leadership, with a strong capacity to mobilise local (and often international) interest and action. However, as discussed in Part II, older forms of community cohesion and decision making have largely declined as a result of urbanisation and the effects of the wider market economy in Europe. Local governance institutions may need reforming from within, to ensure that women, ethnic and user groups are better represented in decision-making structures. Women are often absent from networks of forestry associations and unions in the region,[6] while men are often over-represented in local management institutions, organisations and networks. The characteristics of what constitutes effective local governance are well documented, and include the existence of consensus arrangements; clearly defined user rights, operational rules and sanctions; ability to participate in formulating and modifying rules directly or through elected representatives; an enabling policy framework or state tolerance. It is also generally agreed that the size and constituency of the group is also important, with small numbers of users contributing to greater success of local governance.[7] With democratic and open systems in place, local forest governance can provide a socially legitimate basis for sustainable forest management.

PUBLIC FOREST GOVERNANCE

In general, the social and political pressure for more transparency and accountability in decision making throughout the western world, is prompting changes within public forest agencies and NGOs concerned with forest management and conservation.[8] The democratisation of forest planning, while uneven, is creating new political space, and generating further opportunities for involvement in forest planning, and new forms of public and co-management institutions. As the examples from *Finland* and *Switzerland* illustrated, many institutional reforms and developments have been undertaken, including (re)training of forestry professionals; research, and organisational development to strengthen the relevance of, and give weight to, other forest values and new participatory planning and management procedures. Changes within public forest agencies within Europe have been very uneven to date, and are not without their struggles.

Public forest management has often bequeathed a legacy of structural, legal and economic impediments to wider involvement in forest management. As the study of *French Forest Communes* illustrated, substantial management decisions regarding commune forests still fall under the control of state forest authorities. Effective community involvement in forest management may require more effective devolution of power to the local level. As seen by the national contexts in Part IV, national forest agency capacity to address wider values, including social and ecological functions, is mixed but frequently weak throughout Europe. Some experiences suggest public participatory processes have not altered the basic relationship between the public and forest agency, which continue to be characterised by conflicts in some areas. Failures can result from the technocratic and paternalistic attitudes adopted by some forest agencies, who try to manage forests and forest-related development for communities, rather than establishing responsive partnerships with them. There is also concern that wider participation, where it exists, rarely involves all stakeholders on an equal footing, and is coopted by more powerful interests.[9] Public forest services clearly need further reform and new training opportunities to adapt to new approaches. The participation of women in forestry needs to be actively encouraged through recruiting more women to forestry courses, senior forest appointments, and ensuring equal salaries.[10] New political fora, which encourage a range of partnerships, may be required to develop participatory forest management, rather than a single controlling organisation.[11]

In short, more open, accountable and transparent forms of governance need to be encouraged to counter powerful interest groups which may dominate decision making at local, regional and national levels, and in both private and public sectors. Most governance institutions can be improved to enhance social elements of sustainability- such as equity, representation, accountability and transparency. The diverse values, expectations and objectives of forest interest groups emphasises the urgent need for creating a range of fora for participatory forest management and mutual learning. It also underscores the necessity for new socially oriented forest management skills such as group facilitation, communication and conflict resolution skills. Understanding the role of changing cultural and spiritual values, local institutions, associations, leadership, social capital, and public participatory processes in decision making is not well developed in Europe and would benefit from further research, networking, and advocacy.

CIFM AND BIODIVERSITY CONSERVATION

An important issue within Europe and elsewhere is whether CIFM makes contributions to the conservation of biodiversity and landscapes. As several of the case studies within this profile testify, community involvement in forest management in Europe has often been motivated by environmental concerns – particularly the loss of native woodland. Several initiatives are committed to ecological forest management, such as the use of local tree species and provenances, selective felling and conservation of old growth areas. Others, such as *Crofters Forestry* and the *Borders Forest Trust* in Scotland, are particularly interested in forest restoration. In contrast to large-scale commercial forestry enterprises, communities frequently value a wide range of forest goods and services, and have diverse forest management objectives which can be more sensitively integrated at a local level with benefits for conservation.[12] However, it should not be assumed that CIFM will always lead to better environmental care. Not all communities prioritise biodiversity and nature conservation over other objectives. Some communities, such as the *Crofters in Scotland*, have found that an effective public education programme is required to introduce the wider community to the ecological aspects of sustainable forest management.

FOREST CERTIFICATION

Several of the communities documented in this profile have benefited from FSC certification such as the *Val di Fiemme* in Italy; the *BioRegional Development Group* in London, and the *Saami in Sweden*. The certification of timber and NTFPs creates both opportunities and challenges for communities, and these are discussed more fully elsewhere.[13] The current debate in Europe focuses on the strengths and weaknesses of the Forest Stewardship Council (FSC) and Pan European Forest Certification (PEFC) schemes. In general, certification is thought to provide opportunities to communities by expanding community access to more lucrative markets, allowing local groups a foot in the door of decision making processes, and by supporting the democratisation of trade. In principle, the FSC scheme has an inclusive decision making structure, with equal voting power divided between social, economic and environmental chambers. It can thus bring many stakeholders together on an equal footing, and enhance social equity, as well as promote more ecological forest management. However, the FSC also presents many challenges, notably the costs of certification for small producers; the exclusion of some local groups, and the lack of transparency of decision-making procedures. Small forest owner associations, in particular, have challenged the inappropriate nature of the FSC in the European context characterised by small, fragmented holdings, and have launched their own PEFC scheme. This in turn has been widely critiqued by environmental and social NGOs alike for failing to develop overall performance-based criteria; omitting to put in place clear control mechanisms to ensure that certified timber does come from well-managed forests; failure to fully include all stakeholders in decision-making processes, including indigenous peoples, and for focusing only on Europe.[14]

ENABLING POLICIES

Are changing policies providing more opportunities for new patterns of CIFM in Europe? Since UNCED and other international policy initiative,[15] far more attention has been given to sustainable forestry, and the role of community initiatives and public participation in forest planning and management. These policies contribute to the evolution of a common international agenda; help establish basic levels of accepted best prac-

tices, and can also generate peer pressure among governments to take action. Since the early 1990s, most European countries have developed multiple use forest policies – which recognise the ecological, social and economic roles of forests; and are developing various criteria and indicators for sustainable forest management. As discussed in the national contexts in Part IV many of these policies appear to be supportive of participatory processes and greater community involvement in forest management.

Opportunities exist for groups, and sometimes individuals, to participate in policy decision-making fora, and to promote community involvement in sustainable forest management within Europe at all policy levels, including:

◆*International Fora*, such as the UN Forum on Forests (UNFF) – which is a continuing development of several previous International Fora on Forests. Multi-stakeholder dialogues are proposed for the UNFF as a means of enhancing NGO, indigenous peoples and other major group participation in forest decision making. At international levels, it is increasingly clear that governments 'can't do it alone' and depend on public-private partnerships. However, it remains to be seen whether multi-stakeholder dialogues are adopted by the UN Forum on Forests, and how effectively community perspectives are integrated into these discussions.

◆*Regional Fora,* such as the Inter -Ministerial Conferences on Protection of Forests in Europe.

◆*Regional Policy Networks,* such as the working group on 'Mountain Forests, Peoples and Communities', which is run from the European Observatory of Mountain Forests, based in France.

◆*National and Local Advisory Panels,* such as the Panel on 'Forestry for People' in Scotland chaired by the UK's Forestry Commission.

POWER AND THE POLICY PROCESS

Despite these international and national policy developments, it is also apparent that there is considerable resistance to change. A close examination of the policy process reveals the political nature of policy change and implementation, which may be considered the products of negotiations of many different actors of unequal power relations. At the international level, it is widely appreciated that policy development is frequently constrained by those most threatened by change, and that regulation of international policies is problematic because there are often no clear enforcement mechanisms. In some cases, NGOs have played a key role in the creation of soft law instruments such as the Forest Stewardship Council's Principles and Criteria for Natural Resource Management, partly because international policy regimes are perceived as ineffective responses to both general and specific problems.

At national level, most countries are faced with the legacy of an earlier industrial, technocratic forestry paradigm, which, until relatively recently, emphasised timber production objectives, to serve economic growth and national development goals, at the expense of other social, economic and ecological values. The transition to a new sustainability paradigm is uneven within and between countries, and is faced with many contradictions between ideals and practice. Moreover, the question of whether international and national policies can exert effective influence over increasingly powerful corporate interests, is an emerging issue of global significance.

At local levels, several of the case studies indicated that communities, such as those in *Val di Fiemme* in Italy, have had to struggle against government pressure to retain access to their forest resources, and their traditional rights to control them. In the case of the *baldios* in Portugal these struggles continue. In other cases, some countries resist ratifying international conventions that support local and indigenous peoples' rights. Sweden is one of the few European countries not to have signed the ILO Human Rights Convention 169 which would support the Saami's traditional usufruct rights in private forests. Analysis of successful policy change for CIFM at local levels, such as *Crofter forestry* in Scotland, demonstrates how the sustained and strategic political action of coalitions of local and non-local groups is critical in effecting policy changes.

In short, despite the emerging international and national 'enabling policy context', decision making within the forest sector continues to be influenced by various layered and intersecting structures of power. Further progress towards holistic, decentralised, multiple-objective management which supports CIFM is needed within Europe. This progress will depend on improved media treatment; greater independence for forest research; greater transparency of public forest agency decision making, further electoral reforms in some countries; and effective political action at the grassroots.

LESSONS FOR GOVERNMENTS

◆*Benefits of CIFM*. A review of CIFM in Europe indicates that rural and urban people can make significant contributions to sustainable forest management. With appropriate institutional support, communities can successfully undertake forest regeneration, management and protection, providing diverse benefits to a wide range of actors, see Contributions of CIFM to Sustainability in the next chapter.

◆*Diverse Forms of CIFM*. There is no one model of CIFM in Europe. This profile recognises and learns from a wide variety of institutional arrangements for community involvement in forest management in the region, such as:

– older, but progressive forms of communal forest ownership and management: *Val di Fiemme in Italy; the baldios in Portugal; the Swedish Forest Commons*

– contemporary, self-mobilised community forestry initiatives in rural and urban areas: *the Crofters in Scotland; the Borders Forest Trust; The BioRegional Development Group*

– commune forestry: *French Forest Communes*

– forest consortia: *Italian Forest Consortia*

– multistakeholder-based resource management associations: *the pastoral restoration groups described in Sixt, France*

– local associations of small private forest owners and producers with collective interests: *European Small Forest Owner Associations*

– problems encountered by indigenous peoples in forest management: *the Saami in Sweden*

– forest worker unions

– public forest agencies which encourage greater public participation in forest planning and management: *Examples from Finland, Denmark and Switzerland.*

The diverse forms of CIFM are rooted in western Europe's particular political economy, and can provide useful lessons for other regions, including those with communal land tenure systems; private forest ownership and co-management systems. Supporting different approaches allows for balance and flexibility in the face of social diversity and change.

◆*Government Support Required.* Although community forestry initiatives may evolve without direct government intervention or assistance, most rely on some kind of Government support or co-management arrangement. Governments play key roles in CIFM by providing supportive policy, legal and economic frameworks, technical aid and communications. While communities require political autonomy and freedom to operate, they require stable and flexible legislation, including secure ownership, access and usufruct rights; the freedom to organise into groups and so on. Governments can also provide important incentives for CIFM. Direct financial incentives can be in cash (e.g. grants) or kind (e.g. the provision of office equipment, training, roads). Indirect fiscal incentives, such as tax reductions on land or income also underpin some forms of CIFM. Public authorities can also provide extension services, helping communities acquire new practical skills for land use changes. Public forest administrations may also initiate or facilitate community forestry initiatives or be involved in co-management arrangements, such as forest consortia. Governments also have a communications role, by raising awareness about forests and CIFM through the media, public education programmes, teacher training, school curricula and so on.

◆*Redressing the Balance.* History illustrates that the changes associated with establishment of modern nation states, the intensification of agriculture, urbanization and the evolution of the industrial forestry paradigm, has resulted in the de-linking of people from forests, and forests from agriculture, and has undermined many community institutions. Governments can play a key role in compensating for historical inequities. They can support new opportunities for CIFM in Europe through:

– redistribution of land rights to allow greater community access and use of forests

– further policy reform

– new economic incentives for poorer groups and sustainable practices

– political support of marginalised communities in the context of vested private interests

– more devolution of power to local levels

– reorganisation of forest administrations and state forest authorities

– the development of new market opportunities for wood and non-wood products

– support of small scale wood processing

– extension services.

LESSONS FOR NGOS

◆*The Role of CIFM.* Social and environmental NGOs increasingly recognise that classic models of rural development and biodiversity conservation, including exclusionary protected area systems, are frequently socially inappropriate and often fail to achieve their objectives. Community based projects exhibit a number of compelling features attractive to environmental and social NGOs alike (see *Contributions of CIFM to Sustainability* in the next chapter). Community initiatives are able to integrate complex social, economic and environmental values and objectives in ways that specialised NGO groups may find hard to justify and pursue.

Self-mobilised communities are usually highly motivated, have strong local leadership, and are generally socially accountable to a wide range of local people. They are often open to, and adopt progressive policies and practices. Local forest user groups and forest workers are also in a good position to monitor sustainable practices and can make important contributions towards monitoring activities. CIFM can thus make important contributions to sustainability.

◆*NGO Support.* Local NGOs can sometimes play a key role in representing, supporting and sometimes catalyzing CIFM. They play an important role in helping mainstream, at local and other governance levels, international best practices such as:

– participatory processes

– gender equity

– conflict resolution

– environmental awareness

– promoting ecologically sensitive management and technologies

– building partnerships and coalitions for change

– demystifying political structures and processes

– lobbying governments, businesses and others for policy and legislative changes

– providing incentives in cash and kind: equipment, infrastructure, management training

– providing organisational and technical skills and marketing assistance

– playing a key role in research, analysis, education

– acting as 'watch dogs' – roles which may be beyond the capacities of communities to undertake themselves

– providing an important external voice for communities

– playing a key role in promoting CIFM in various policy fora.

◆*Community Differences.* It is important for outside groups supporting CIFM to recognise and address social differences within 'communities', which shape access to resources. Communities are not homogenous groups, but can be differentiated along many axes such as gender, age, ethnicity, cultural values, access to, and control of, resources like land, labour, capital, information and so on – all of which have a profound influence on those who benefit from and who bear the costs of forestry activities. Problems faced by women in forest management are frequently different to those faced by men. Participation of forest workers in forestry decision making is essential for ensuring that issues of workers' health, safety and equity are factored into decision making. NGOs also need to be flexible and identify and support a diversity of local organisational types.

LESSONS FOR RURAL AND URBAN COMMUNITIES

◆*Local Benefits of CIFM.* Local involvement in tree and forest management can produce many economic, cultural and ecological benefits in rural and urban areas (see *Contributions of CIFM to Sustainability* in the next chapter). CIFM can be a way of re-establishing local access and usufruct rights in woodlands and forests, and resisting trends towards greater state or private control. Woodland establishment, tree management and wood processing can generate local livelihood and employment opportunities, and provide local products for local needs. It can help protect and restore landscapes, and produce a sense of cultural identity and pride in local achievements.

◆*Adapting to Change.* Evidence from the European case studies indicates that CIFM has thrived where a viable economic base is supported by common ethical and cultural values. However, in order to remain viable and robust, successful communities have been flexible and open to new opportunities, and adopted many progressive policies and practices including:

– new silvicultural techniques

– 'green' management systems

– upgrading and adopting new processing technologies

– supplying new timber and non-timber products

– incorporating quality control mechanisms for sophisticated markets

– using alternative patterns of sales and trade

– developing new organisational arrangements

– finding new sources of finance

– initiating partnerships with other local and non-local groups

– developing communications strategies.

In general, it is clear that finding market solutions for local products generates real incentives for communities to invest and reinvest in forests, and enhances their ability to adapt to changing social and economic conditions.

◆*Local institutions.* Traditional management institutions often form appropriate foundations for new organisational arrangements, but may require reform to make them more equitable and learning oriented. Democratic assemblies usually prove effective instruments for finding locally acceptable compromises in relation to local resource use, and for curbing the effects of local vested interests. Effective and sustainable CIFM frequently involves on-going efforts to overcome factional differences within communities. Associations of small private owners can overcome the disadvantages of small individual holdings, and be effective bodies for sharing costs of infrastructure, negotiating better prices for products and equipment, and so on.

◆*The Role of Local Leaders.* The review process indicated the key role of local leaders in CIFM.

They provide and articulate a vision for CIFM, and are influential in raising local and wider awareness. They attract, catalyse, persuade, build trust, and help organise sympathetic community members, and act as a focal point for wider groups. They also play an important role in networking with other groups. Experience suggests that leadership skills change over time. While awareness raising, advocacy and diplomacy skills are particularly important in early stages, management and technical skills increase in importance as projects develop. It is clear that a mix of leaders with different skills is important to CIFM.

◆*Coalitions for Change.* Local people can be very effective in catalyzing institutional changes for greater CIFM in Europe. However, it is widely appreciated that local communities can not always do it on their own, but rely on the support of public figureheads, networks, partnerships and coalitions of local, national and sometimes international groups within government and NGO sectors. Many communities have realised that government bodies and NGOs are not monolithic entities, but made up of individuals and groups who often share similar values, ideas and visions. It is possible to have direct contact with influential public figures, and forge strategic coalitions with sympathetic individuals and external groups to help shift the balance of power necessary for policy and legislative changes in favour of CIFM. Other organisations provide an important channel for voicing local community concerns in various policy fora, helping to reinforce public support. Partnerships also provide local groups with information about markets, community organisation, policies, and financial resources.

◆*Step Wise Approach.* CIFM is a process of evolution and adaptation to shifting opportunities and challenges. While it is important to have an overall vision for local involvement in forest management, objectives are usually only realised in a strategic step-by-step approach.

NOTES

[1] See Ministerial Conference on the Protection of Forests in Europe, Liaison Unit, Vienna (2000): *The Role of Forests and Forestry in Rural Development – Implications for Forest Policy. Proceedings on the International Seminar, 5-7 July 2000, Vienna, Austria.*

[2] Jeanrenaud, S. & Jeanrenaud, J.P.(1997): *Thinking Politically about Community Forestry and Biodiversity: Insider-driven initiatives in Scotland. Rural Development* Network Paper 20c, Winter 1996/7. London: Overseas Development Institute.

[3] There are well known exceptions to this rule. Secure property rights do not always underpin sustainable forest management. People may prefer alternative land uses, such as intensive agriculture.

[4] Lynch,O. J. & Alcorn, J. (1994): "Tenurial Rights and Community-Based Conservation". In Western, D. & Wright, R.M. (Eds)(1994): *Natural Connections: Perspectives in Community Based Conservation.* Washington DC: Island Press

[5] There are some important qualifications to this rule. In mountain regions in particular, land holdings – even among private forest owners – are not always clearly demarcated.

[6] Faugère, I. (1999): *The Role of Women on Forest Properties in Haute-Savoie: Initial Researches.* Geneva Timber and Forest Discussion Papers No.13. New York and Geneva: United Nations.

[7] Hobley, M. & Shah, K. (1996): "What makes a Local Organisation Robust?" ODI Resource Perspectives No.11. London: Overseas Development Institute.

[8] ILO (1997): *People, Forests and Sustainability. Social Elements of Sustainable Forest Management in Europe.* Geneva: International Labour Organisation.. Also, Kennedy, J.J. Dombeck, M.P. & Koch, N.E. (1998): "Values, beliefs and management of public forests in the Western world at the close of the twentieth century." In UNASYLVA 49 (192): 16-26 .

[9] Hildyard, N. Hedge, P. Wolvekamp & Reddy, S. (1998): "Same platform, different train: the politics of participation". In UNASYLVA 49(194): 26-34.

[10] Ekberg, K. (1997): "How to Increase the Participation of Women in Forestry – Ideas and Ongoing Work". In ILO (1997) op.cit.

[11] Henderson, D. & Krahl, L. (1996): "Public Management of Federal Forest Land in the United States".Unasylva No 47 (184): 55-61.

[12] Ghimire, K. & Pimbert, M. (1997): *Social Change and Conservation.* London: Earthscan.

[13] See Forest, Trees and People Newsletter, No.43 November 2000, which is dedicated to certification and community forestry.

[14] Ozinga, S. (2000): *The European NGO position on the Pan European Forest Certification Scheme in Forest,* Trees and People Newsletter No.43, November 2000.

[15] These include: Agenda 21; the Ministerial l Conferences on the Protection of Forests in Europe; the Environment for Europe Process; the EU Forest Strategy and EU Agricultural Policy; the EU CAP reform; the Forest Stewardship Council, etc. See Appendix 3 for brief descriptions.

Part VII

CONCLUSION AND VISION FOR CIFM IN EUROPE

This chapter briefly examines the diverse contributions of CIFM to sustainability[1] in Europe. It then proposes some recommendations for policy and action, before outlining a vision for CIFM in Europe for the 21st century.

THE CONTRIBUTION OF CIFM IN EUROPE TO SUSTAINABILITY

In this profile sustainability is understood to encompass three interlinked goals of: **Economic Viability, Social Equity** and **Environmental Protection**. The European case studies illustrate how community involvement in forest, woodland and tree management can make important contributions to these goals. Case studies with experience of particular issues are indicated below in italics:

CONTRIBUTIONS TO ECONOMIC VIABILITY

◆Provision of non-marketed products: *Val di Fiemme in Italy; baldios of Portugal; NTFPs in the Mediterranean and Boreal;*

◆Revenue from timber and NTFP trade: *Val di Fiemme in Italy; Swedish Forest Commons; NTFPs in the Mediterranean and Boreal; the BioRegional Development Group in the UK; Woodschool Initiative of the Borders Forest Trust; Small Forest Owners; French Forest Communes;*

◆Revenue from grants: *Crofters in Scotland; Borders Forest Trust; BioRegional Development Group;*

◆Employment: *Forest Workers; Small Forest Owners;*

◆Integral components of agricultural and pastoral systems: *Val di Fiemme in Italy; the baldios of Portugal; Saami in Sweden.*

CONTRIBUTIONS TO SOCIAL EQUITY

◆Redistribution of land rights: *Crofters in Scotland;*

◆Promotion of indigenous peoples' rights: *Saami in Sweden;*

◆Wages and working conditions: *Forest Workers;*

◆Gender equity: *Forest Workers;*

◆Local democratic institutions: *Val di Fiemme in Italy; baldios of Portugal; Crofters in Scotland; Swedish Forest Commons; Small Forest Owners;*

◆Multi-stakeholder dialogue/Mutual learning: *French Forest Communes; Public Forest Agencies in Finland, Switzerland, Denmark; Forest Workers; Forest Consortia in Italy;*

◆Political coalitions for change: *Crofters in Scotland; Saami in Sweden.*

CONTRIBUTIONS TO ENVIRONMENTAL PROTECTION

◆Commitment to local ecology and landscapes: *Val di Fiemme in Italy; Crofters in Scotland;*

Borders Forest Trust; Saami in Sweden; NFTPs in the Mediterranean and Boreal regions; BioRegional Development Group;

◆Ecological forest management: *Borders Forest Trust; NFTPs in the Mediterranean and Boreal regions; Val di Fiemme in Italy; Saami in Sweden; Crofters in Scotland; Small Forest Owners; Forest Workers; BioeRegional Development Group in London;*

◆Ecological restoration: *Crofters in Scotland; Borders Forest Trust; NFTPs in the Mediterranean and Boreal regions; BioRegional Development Group;*

◆Forest certification: *Val di Fiemme in Italy, Saami in Sweden; BioeRegional Development Group in London; Small Forest Owners.*

The studies illustrate how different forms of CIFM integrate economic, social and environmental values in diverse ways and proportions, representing their unique and comparative contributions to sustainability. Their success in contributing to sustainability is defined by the objectives and scope of each initiative, but also by the opportunities and constraints provided by prevailing policies, land ownership patterns, economic contexts and governance structures – and their interlinked effects. These were briefly reviewed in the preceding chapter.

RECOMMENDATIONS FOR GREATER CIFM IN EUROPE

Based on the opportunities and challenges for CIFM, and the lessons in the preceeding chapter, the profile promotes the following recommendations for policy and action in Europe:

◆A diversity of approaches to community involvement in forest management in Europe;

◆Policy reform that emphasises the three interlinked goals of sustainability: economic viability; social equity and environmental protection;

◆Secure forest property and usufruct rights;

◆Participatory approaches to local, co-management and public forest governance;

◆Economic reforms to support sustainable rural and urban livelihoods;

◆Partnerships and coalitions for community involvement in forest management;

◆Forest agency reform;

◆Intersectoral coherence and integration of land use policies, such as farming, forestry, tourism and sustainable rural businesses.

LOOKING AHEAD: A VISION FOR CIFM IN THE 21ST CENTURY

CIFM is seen as an important spearhead of sustainability in post-industrial Europe. It is viewed as a means of benefiting people as well as forests. CIFM is not seen as a retreat to the past, or based on some romanticised view of rural communities living in harmony with nature. Many rural and urban peoples involved in forest management are *sustainability pioneers*, and demonstrate innovative and diverse ways of combining new livelihood opportunities with concerns for social equity, ethics, biodiversity conservation and landscape values. Whilst local, such initiatives also produce benefits of wider significance. CIFM is thus seen as a way of *integrating* complex social, economic, ecological and spiritual dimensions of forest management in dynamic ways, that larger private and public enterprises can find hard to pursue or justify. The three interlinked goals of sustainability – economic viability, social equity and environmental protection – should be at the heart of forest management in the 21st century. However, significant institutional changes are required to realise the great potential of CIFM, which is still under-valued and under-supported in Europe. The challenge for governments and other actors, is to create and support new initiatives which allow people to reconnect with Europe's abundant and often under-utilised forests for a wide range of objectives. This will involve policy reform and greater access to forests; democrastisation of planning and trade; development of appropriate incentives; new processing technologies and markets; enrichment of professional forestry skills, and mutual learning among a range of stakeholders.

Lessons on CIFM from different geographical regions suggest that there may be evidence of a global transition in forest management. It appears that an increasing number of nations in both North and South are developing policies and operational mechanisms to provide much more active roles for local and indigenous peoples in practising sustainable forest management. This transition may be unfolding in different ways and is at different stages of development in each nation.[2] The European experience suggests that caution should be exercised in pinpointing trends in CIFM in the region. While there are hopeful signs of a transition in policies and activities, analysis suggests significant historical and institutional constraints and powerful resistance to greater CIFM by some groups. A transition to further CIFM in the region is not inevitable. The vision of CIFM in providing a vast range of social, economic and environmental benefits to rural and urban peoples, will require long term commitments, strategic political pressure and sustained financial support to evolve. But it is, we believe, a vision worth working for.

Box 13   **VISION FOR CIFM IN 21ST CENTURY EUROPE**

◆CIFM is seen as a way of (re) connecting people with forests in rural and urban areas for sustainable futures.

◆CIFM is viewed as a means of integrating economic, social, cultural, spiritual and ecological values in diverse, innovative, and evolving ways.

◆CIFM is appreciated as a way of encouraging participatory processes, mutual learning, respect, partnerships and coalitions for the benefit of both people and forests.

NOTES

[1] See glossary for definition.

[2] Poffenberger, M. (1996): A Long Term Strategy of the IUCN Working Group on Community Involvement in Forest Management. Gland: IUCN The World Conservation Union.

ABOUT THE CONTRIBUTORS

Nanna Borchert is German. She is a forest ecologist and has worked on international forest issues with focus on protection and sustainable forest management of the boreal forests in the Northern hemisphere. Her main interest is in supporting local/indigenous peoples to manage the forests themselves and work against large scale environmental destruction. During the last few years she was employed by Saami reindeer herding communities in Sweden to coordinate a European wide project with the aim of raising awareness about the existential land right struggle the Saami are facing today.

Jill Bowling is from Australia. She is currently Director of the Global Forestry Programme, International Federation of Building and Wood Workers (IFBWW), based in Switzerland. This programme promotes the participation of workers in sustainable forest management and forest certification activities globally. She has a background Zoology (BSc), and Human Sciences (PhD). In the 1990s she worked in the USA as the Director of the State Forests Programme, Oregon Department of Forestry, and earlier as a Policy Analyst for the Natural Resources Planning Team, in the Oregon Governor's Office.

Roland Brouwer is from the Netherlands. His PhD thesis examined the afforestation of the commons in Portugal by the state Forestry Service, showing that tree planting was part of the subjugation of the local population to state law and power. He has a background in Forest Engineering, specialising in forest policy and management (MSc). He is currently based in Mozambique, working with support from a Dutch NGO called "Dienst Over Grenzen" as an assistant professor at the Faculty of Agronomy and Forestry Engineering, where he teaches and does research into several aspects of forest policy and community natural resource management.

Lars Carlsson is from Sweden and holds a position as associate professor at the Division of Political Science, Luleå University of Technology, Luleå Sweden. He has published a number of articles, book chapters, and papers about policy analysis and the management of common-pool resources. Currently Dr. Carlsson is involved in research for the Forestry Project at the International Institute for Applied Systems Analysis (IIASA), Laxenburg, Austria. The project is interdisciplinary and deals with institutional problems of the Russian forest sector.

Bruno Crosignani is from Italy, and is currently a forest inspector of the Forest Services in the Autonomous Province of Trento in Italy. He has worked mainly in the management of common forests; on the stability of mountain terrain, and on building mountain forest roads. Lately he has worked on eco-certification of saw-mills and on safety and environmental management systems.

Andréa Finger is German. She is based in France, and is currently working on a PhD on participation in local forest management in the Alps. She has worked with the United Nations Research Institute on Social Development (UNRISD) on the social impacts of protected areas; as a forest policy analyst with IUCN and WWF and lately as a coordinator and co-reporter of the ECE/ILO/FAO Team of Specialists on Public Participation in Forestry. She has a background in Forest Resources Management (MSc).

Mandy Haggith and **Bill Ritchie** live on a woodland croft in Assynt, on the west coast of Scotland. Mandy studied philosophy and used to work in academia. Bill studied law and used to work for the government. Their mutual passion for forests brought them together to set up 'worldforests' in 1997, a small organisation dedicated to 'research, writing and revolution in support of people in forests'. They are the editors of 'Taiga News', the newsletter of the Taiga Rescue Network. Their current research work is with an interdisciplinary team based at CIFOR, Zimbabwe, modeling a social forest landscape as a visioning tool for policy makers. Their revolutionary activities include photography, poetry and politics.

Yves Kazemi is a Swiss forest engineer working as forestry consultant and forest policy analyst - at national and international level. His main field of activities are the social aspects of forestry and the participatory approaches to natural resources and forest management. He co-ordinated both the FAO/ECE/ILO Team of Specialists on Social Aspects of Forestry (1998) and the Team on Public Participation in Forestry (2000). He has background in Forest Engineering (BSc) and Public Administration (MA). He has worked on several research programs dealing with social attitudes and perceptions towards nature and forests.

Sally Jeanrenaud is British, and is currently working as an independent researcher and consultant on development/conservation issues, based in Switzerland. She has worked in Nepal, Rwanda, and Cameroon in rural development forestry and conservation projects. She has a background in Development Studies (BA Hons), specialising in social issues, Forestry and Land Management (MSc) and Development Studies and Conservation (PhD).

Olaaf Johansonn is a Saami reindeer herder in Northern Sweden. He is also a politician and activist working towards nature protection as well as towards securing Saami rights. He often represents the Saami at international meetings on questions about forests, biodiversity and indigenous peoples.

Simon Levy is British, based in south London. He is currently the Project Manager for the Urban Forestry Project of the Bioregional Development Group. He also works as an assessor for the Forest Stewardship Council (FSC) certification. He has worked in forestry for the past six years and developed the urban forestry project and some early certification methodologies. He has a background in Environmental Management (MSc).

Sarah Lloyd is from the United States. Her Masters degree thesis in Rural Development Studies, focused on relationships between socio-economic and ecological space in a traditionally natural resource-based community in the north of Sweden. She has been active in the Taiga Rescue Network (TRN) for many years, working for the protection and sustainable use of the boreal forests. She has worked on development of sustainable non-timber forest products (NTFPs) in the Russian Far East and conducted research on NTFPs in the Nordic countries. She has recently moved to the ancestral Lloyd farm in rural Wisconsin, USA to start a small farm enterprise and will continue work, both privately and professionally, on sustainable rural development.

Willie McGhee is Scottish. He works as a Woodland Co-ordinator for the Borders Forest Trust, based in Ancrum, Scotland. He has worked in forestry and forest ecology for 10 years, and is a Trustee of the Millennium Forest for Scotland Trust, a Director of the Southern Uplands Partnership, Chairman of Habitat Scotland and a member of the Forestry Commissions Forests and People Advisory Panel. He is a graduate of Edinburgh University with a degree in Ecology (Hons Forestry)

Yorgos Moussouris is Greek. He has worked with WWF on issues pertinent to the importance of NTFPs in rural development in the Mediterranean. Recently he worked as a consultant on environmental planning for Alcyon, an Athens based firm. He has a background in engineering, geology and environmental planning (Msc). He currently works for the Ministry of National Economy in project management in Greece.

Pedro Regato is Spanish. He coordinates the Forest Unit of the WWF Mediterranean Programme, based in Rome. This unit addresses forest diversity and conservation; sustainable use and forest management, related to NTFPs, forest restoration and forest fires in the Mediterranean Region; as well as training and education activities for Mediterranean NGOs, forest managers, protected areas managers, and researchers. He has a background in biology (MSc) and Forest Ecology (PhD), and many years of research experience in ecological and rural-land use issues in southern Europe.

Maria João A. Pereira is Portuguese. She is currently engaged by the Forest Authority as an external consultant for the legal and institutional framework for the implementation of the Portuguese Forest Act 1996, dealing with market-driven instruments for SFM, such as insurance, fiscal regulation and funding. She was in the taskforce for the Lisbon Conference- the Third Ministerial Conference on the Protection of Forests in Europe (Lisbon 1998). She has a background in Forestry, and a post-graduate diploma in Rural Development. She has post-graduate experience in Economics and Econometrics, and teaches mathematics for economists.

Pauli Wallenius is Finnish. He works for Metsähallitus (Finnish Forest and Park Service), where he is responsible for forest management, silviculture and planning. During 1990s he was responsible for developing public participation practises in Metsähallitus. He has studied in USA and worked together with US Forest Service in developing participatory processes. He was also a member of the FAO/ECE/ILO international specialist group which developed principles of public participation in European forestry.

Pier Carlo Zingari is Italian but has been based in France since the early 1990s. He initiated, together with representatives of forest communities, the European Observatory of Mountain Forest (EOMF) in Savoy, France. The EOMF aims to bring together forest and non-forest peoples' views, experiences and interests at the different levels in the conservation and sustainable management of fragile mountain ecosystems. He has a background in Forestry (BSc), and Forest Economics and Planning (PhD), and research experience on the ecology and management of mountain silvo-pastoral systems. He has also worked in Costa Rica with FAO.

GLOSSARY OF TERMS

Coalitions

Groups or individuals who deliberately collaborate or 'co-align' to effect greater changes than could be achieved than acting alone. Together, groups exploit a 'middle ground' of shared interests.

Historical analyses of community forestry in several regions has indicated the influential role of informal social networks of local, national and international actors in promoting new policies and legislation, securing funding and in underpinning the success of many community forestry initiatives.[1] Recognition of the effects of unequal power relations between stakeholders and interest groups increasingly prompts analysis and support of coalitions in securing benefits.

Co-Management (short for collaborative or joint management)

This term has been defined as "...durable, verifiable and equitable forms of participation, involving all relevant and legitimate stakeholders in the management and conservation of resources".[2]

Common (or communal) Property Forest

Forest land owned by a group (co-owners) who hold exclusive rights and share duties towards that resource. It is perhaps better understood as a group-owned private forest. This type of property exists in several parts of Europe, such as Italy, Sweden, Portugal and Switzerland. Communal forest properties have several advantages. They keep large resource systems in tact without having to divide it into pieces, which is useful when a resource is large and difficult to demarcate. They help ensure productive efficiency through the internalisation of externalities, for example – where people make decisions in a water catchment area. They also address the concerns of more marginal farmers whose livelihoods may depend upon shared resources not available under individual private or state property. In addition to public and individual private ownership, communal forest properties can provide a socially equitable and environmentally sound basis for forest management.

Commune

The smallest administrative unit in countries such as France and Switzerland. They are locally self-governed by a Mayor and municipal council. It is important to distinguish *commune forests* from common (or communal) property forests, see above.

Community

In this profile the term refers to *communities of interest*, which may include: property, user and access rights; livelihoods based on the production of timber and non timber products; employment; cultural identity; leisure and recreation; biodiversity conservation and ecological restoration. See Part I for more details.

Consortium

A partnership. For example, forest consortia in Italy are legal management entities consisting of public and private forest owners.

Decentralisation

The transfer of authority and responsibility for public functions from the central government to the subordinate government and/or private sector. It includes political, administrative, fiscal and market dimensions.

Democracy

Government by the people. Power is exercised directly or indirectly through a system of representation and delegated authority which is periodically renewed.

Democratisation

Affords citizens and their representatives more influence in formulating and implementing policies. It may be considered a form of political decentralisation. It is based on the belief that decisions made with greater participation will be better informed and more relevant to the diverse interests of society than those made only by national political authorities.

Governance

In the forest sector, this is literally the act of exercising control over forest resources. It implies various processes, organisational structures, and types of behaviour which shape forest resource use and protection. Governance patterns are interesting partly because they reflect different attitudes towards representation, accountability, rights, responsibilities and equity – key social elements of sustainability.

Institutions

May be understood as regularised patterns of social behaviour which emerge from underlying rules, values, norms and dominant worldviews. Institutions may be formal (the rule of law) or informal (social norms). They subsist independently of individuals and usually escape the control of any one individual. Institutions are also recursive, in that they are made and remade through peoples' practices. Institutions mediate access to resources at various levels.

Incentives

Economic incentives can take a large number of forms and can be categorised in several ways, see Part IV.

Direct incentives are applied to achieve specific objectives, and can either be in cash or in kind. Cash incentives include grants, subsidies, loans, fees, compensation, etc. Direct incentives are usually popular because they can be used in flexible and individual ways. But, they can create dependency on outside aid. Direct incentives in kind include material goods offered to institutions, communities or individuals in return for their contributions, such as equipment, land, access to resources, etc.

Indirect incentives involve the application of fiscal, service and social policies to promote investment, production and employment in community involvement in forest management. Fiscal incentives are a legal and statutory means of channelling funds towards preferred activities. They include tax exemptions or allowances, insurance, price supports, guarantees etc. Tax incentives allow individuals or communities to be wholly or partially exempted from government taxes (on land, income, sales, inheritance, or capital) in return for particular activities. Perverse incentives including taxes encourage undesirable behaviour.

Leaseholds

Contracts which permit tenants to benefit from the land for a specified period of time, usually for rent. Forest land belonging to private or public owners is sometimes leased to individuals or communities for forestry purposes.[3]

Livelihoods

Livelihoods comprise the capabilities, assets (including both material and social resources) and activities required for a means of living. A livelihood is sustainable when it can cope with and recover from stresses and shocks and maintain or enhance its capacities and assets both now and in the future, while not undermining the natural resource base.[4,5]

Municipality

A district (village, town or city) having powers of self-government

Non-Timber Forest Products (NTFPs)

NTFPs may be defined as biological resources of plant and animal origin, harvested from natural forests as well as plantations. They are gathered in the wild, produced as (semi-) domesticated plants in plantations and agroforestry systems, or in intermediate production systems which reflect several degrees of domestication.

Participation

A process through which stakeholders have the potential to influence and share control over initiatives and the decisions and resource which affect them, see Part I for more details.

Power

In simple terms, the ability to act to produce an effect, but there are numerous models and definitions of social power.[6] This profile frequently implies a 'decentred' notion of power (cf. Foucault 1980),[7] which views it as an aspect of all relationships, including non-political ones, and which is expressed throughout society in multiple arenas – from international to local fora. This conception identifies the constant articulation of power and knowledge, and how knowledge is constructed in the service of power. It recognises that power is derived from many different sources – discourses, institutions, social actors, and events. These interact to institutionalise dominant social power/knowledge configurations. Based on this perspective, power is not a 'thing' or substance which can be possessed or given away, rather it is more of a strategic relation. Individuals may be seen as vehicles of the circulation of power, as well as its point of application. Power is mediated through social alignments. This notion of power differs significantly from more traditional notions of class or sovereign power, or those concerned with aspects of personal development.

Post Industrial / Post Modern

Terms used to suggest a cultural condition of going beyond the dominant industrial and modern world views of the 19th and 20th centuries, which are widely percieved as inadequate on social and ecological grounds. While the prefix 'post' suggests that older world views have been surpassed by new knowledges and values, the terms identify themselves by what they are not, rather than specifying and characterising a new era.

Private Ownership of Forests

Forest or other wooded areas owned by individuals, families, cooperatives and corporations which may be involved in agriculture or other operations as well as forestry; private forest enterprises and industries; private corporations and other institutions, such as religious and educational institutions; pension and investment funds; nature conservation societies.[8]

Public Ownership of Forests

This is defined as forest and other wooded land owned by national, state and regional governments, or by government-owned corporations; Crown forest and other wooded land.[9]

Social Capital

Consists, in a narrow sense, of social networks and associated norms that have an effect on the productivity of the community. It is rooted in trust, and is that which facilitates cooperation and coordination for the mutual benefit of members of the group. In a broader sense the term captures vertical as well as horizontal associations, between communities and other groups such as forest agencies, forest certification groups, municipal councils, and so on.

Stakeholder[10]

Any person, group or institution that has an interest in forestry activities. Stakeholders include both intended beneficiaries and intermediaries, and those included or excluded from decision-making processes. They may be divided into two broad groups: the primary stakeholder (local on-site users) who expects to benefit from or be adversely affected by interventions; and secondary stakeholders (such as the wider or even international community) who may play an intermediary role.

The concept of stakeholders emerged partly in response to problems with the notion of 'community', and is useful for extending thinking about the plurality of interests in forest management. However, the concept is limited when it fails to acknowledge the effects of asymmetrical power relations between different groups. Processes of stakeholder participation can become stages on which power relations are played out anew.

Sustainability

There are numerous definitions of sustainability. This profile uses the concept to encompass three interlinked goals: economic viability; social equity and environmental protection. Some of the *social issues* include: poverty alleviation; human rights; wages and working conditions; impact on indigenous peoples; gender equity; employment creation; trust and social capital; representation; accountability; transparency. Some of the *economic issues* include: pricing and profit margins; demand for products and services. Some of the *environmental issues* include: conservation of biodiversity – genes, species, ecosystems; energy consumption; forest quality.

Trusts

Organisations consisting of a number of members, acting together by mutual arrangement, often under a contract, governed by *Trustees*, who are legally entrusted to hold and manage property or funds on behalf of others, see case study on *Community Forestry in the Borders Region of Scotland.*

Usufruct

A term derived from Roman and Civil Law and involves the right to use and enjoy the products of a private or public property, without impairing its substance, see case study on *Saami Reindeer Herders and Forests in Northern Sweden.*

NOTES

[1] Silva, E. (1994): "Thinking Politically about Sustainable Development in the Tropical Forests in Latin America". *Development and Change* 25 (4):699-721. Also Bandyopadhyay, J. (1992): "From Environmental Conflicts to Sustainable Mountain Transformation: Ecological Action in the Garhwal Himalaya". In Ghai, D. & Vivian, J. (Eds): *Grassroots Environmental Action: People's Participation in Sustainable Development.* London: Routledge.

[2] Renard,Y. (1997): "Collaborative Management for Conservation", in Borrini-Feyerabend, G.(Ed) (1997): *Beyond Fences. Seeking Social Sustainability in Conservation.* Vol. 2. A Resource Book. Gland, Switzerland: IUCN.

[3] The Laggan Forestry Initiative in Scotland is a community forestry project using forests leased from the UK Forestry Comission.

[4] Definition adapted from Chambers, R. & Conway, G. (1992): *Sustainable Rural Livelihoods: Practical Concepts for the 21st Century.* IDS Discussion Paper 296. Brighton: Institute of Development Studies.

[5] For an excellent exploration of livelihoods, see Neefjes, K. (2000): *Environments and Livelihoods. Strategies for Sustainability.* Oxford: Oxfam.

[6] See Rowlands, J. (1992) in Nelson, N, & Wright, S. (1997): *Power and Participatory Development.* London: Intermediate Technology Publications.

[7] Foucault, M. (1980): in Gordon, C. (Ed)(1980): *Michel Foucault: Power/Knowledge. Selected Interviews and other Writings 1972-1977.* Brighton: Harvester Press.

[8] UN-ECE/FAO (2000): *Forest Resources of Europe, CIS, North America, Australia, Japan and New Zealand.* New York and Geneva: United Nations.

[9] UN-ECE/FAO (2000): op.cit.

[10] ODA (1995): *Enhancing Stakeholder Participation in Aid Activities.* Technical Note No.13. London: Overseas Development Administration.

APPENDIX 1
PERCENTAGE OF MANAGED FOREST AVAILABLE FOR WOOD SUPPLY IN PRIVATE OWNERSHIP[1]

Country	Owned by Individuals %	Owned by Forest Industries %	Owned by Private Institutions[2] %	Total (1000 ha)
Austria	83	0	0	2,802
Belgium	97	0	3	364
Denmark	67	0	33	301
Finland	79	13	8	12,277
France	84	0	16	10,683
Germany	100	0	0	4,736
Greece	65	0	35	289
Ireland	100	0	0	166
Italy	100	0	0	7
Luxembourg	0	0	0	0
Netherlands	42	0	58	154
Norway	88	5	7	4,800
Portugal	12	88	0	228
Spain	NA	NA	NA	7,501
Sweden	56	44	0	19,332
Switzerland	89	0	11	356
United Kingdom	70	2	28	1,143

[1] UN-ECE/FAO (2000): *Forest Resources of Europe, CIS, North America, Australia, Japan and New Zealand.* New York and Geneva: United Nations.

[2] In some countries NGOs are also important forest landowners, such as conservation organisations in Italy, Netherlands, and the UK. Institutions such as ecclesiastical bodies, academic institutions, financial institutions such as pension trusts also own forests.

APPENDIX 2
PERCENTAGE OF MANAGED FOREST AVAILABLE FOR WOOD SUPPLY IN PUBLIC OWNERSHIP[1]

Country	Total (1000ha)	State Ownership %	Other Public Institutions %
Austria	550	84	16
Belgium	276	25	75
Denmark	140	84	16
Finland	5,164	100	0
France	3,787	39	61
Germany	5,460	62	38
Greece	2,404	85	15
Ireland	380	100	0
Italy	2,044	18	82
Luxembourg	40	25	75
Netherlands	160	72	28
Norway	853	78	22
Portugal	200	13	87
Spain	1,931	2	98
Sweden	1,904	21	79
Switzerland	700	1	99
United Kingdom	965	92	8

[1] UN-ECE/FAO (2000): *Forest Resources of Europe, CIS, North America, Australia, Japan and New Zealand.* New York and Geneva: United Nations.

APPENDIX 3
SOME INTERNATIONAL POLICIES SUPPORTING CIFM

Agenda 21 (1992)

Agenda 21 was the principle outcome of the UN Conference on Environment and Development. It presents a set of integrated strategies and programmes, many of which support community involvement in sustainable and equitable development, including forest management. Several of its chapters address the need to strengthen the role of particular groups such as women, children and youth, indigenous peoples and their communities; workers and their trade unions, etc. Agenda 21 has inspired many community oriented projects at a local authority level, but on the whole, implementation has been very variable throughout Europe.

Convention on Biological Diversity, 1992. 31 ILM 818 (1992), in force 1995

Article 8(j) of the Convention mandates the protection of traditional knowledge, innovations and practices of indigenous and other local communities in the protection of biological diversity, and encourages the equitable sharing of benefits arising from the utilisation of such knowledge, innovations and practices.

Non-Legally Binding Authoritative Statement of Principles for a Global Consensus on the Management, Conservation and Sustainable Development of All Types of Forest, 1992, 31 ILM 881 (1992)

The UNCED Forest Principles contain several principles which support community involvement in forest management. For example, 2(d) states that "Governments should promote and provide opportunities for the participation of interested parties, including local communities and indigenous people, industries, labour, non-government organisations and individuals, forest dwellers and , in the development, implementation and planning of national forest policies", and 12(d) that "Appropriate indigenous capacity and local knowledge regarding the conservation and sustainable development of forests should....be recognised, respected, recorded, developed and, as appropriate, introduced into the implementation of programmes."

ILO Convention No.169 Concerning Indigenous and Tribal Peoples in Independent Countries, 1989, 28 ILM 1077 (1990), in force

ILO Convention No.169 is a leading international human rights instrument calling for the recognition of indigenous territorial rights. Article 13 specifies that references in the Convention to land "shall include the concept of territories, which covers the total environment of the areas which the people concerned occupy or otherwise use". Article 14 mandates recognition of the rights of indigenous peoples to own and possess their traditional territories. Article 15 adds that "The rights of the peoples concerned to natural resources pertaining to their lands shall be specially safeguarded. These rights include the rights of these peoples to participate in the use, management and conservation of these resources". ILO members are legally obliged under Article 19 of the organisation's founding charter to implement the convention. This convention has yet to be effectively enforced, and has not been signed by Sweden because of the Saami issue.

Forest Stewardship Council's Principles and Criteria (1999)

The Forest Stewardship Council (FSC) is a voluntary organisation composed of environmental NGOs, private businesses and human rights groups. It was established to promote the sustainable use of forest resources based on a global standard of ten recognised principles. The FSC is based on the belief that consumers wish to make environmentally informed purchases of forest products. The FSC accredits organisations that have demonstrated capacity to certify whether or not forests are managed according to its 10 principles. It ensures an independent evaluation of a forest management practices. Principle 3 provides that "The legal and customary rights of indigenous peoples to own, use and manage their lands, territories and resources shall be recognised and protected."

As of November 1998, some 7.92 million ha of forests, on 49 sites, had been certified by FSC accredited bodies within Europe.

Forests Certified by FSC in Europe 1998		
Europe	**Million Ha**	**No. of Sites**
Belgium	0.01	3
Czech Republic	0.01	1
Germany	0.00	1
Italy	0.01	1
Netherlands	0.05	8
Poland	2.24	5
Sweden	5.32	16
Switzerland	0.00	2
United Kingdom	0.02	12
TOTAL	7.92	49

Source: Forest Stewardship Council, Oaxaca, Mexico.

Community experience with FSC certification in Europe is mixed. On the one hand it has provided a voice to some forest users – such as the Saami – for the first time. On the other, other communities have objected to the FSC because they have been left outside forest certification processes. Some European small forest owner associations have also rejected FSC standards, because it is seen as inappropriate to European conditions, see Part VI.

European (Aarhus) Convention on Access to Information, Public Participation in Decision making and Access to Justice in Environmental Matters (1998), not yet in force.

The Aarhus Convention grows out of an international process to define the concept of public participation in the context of sustainable development. The three principles of the Convention, broadly stated, are: 1) the public should have access to environmental information, with limited explicit exceptions (the principle of access to information); 2) the public should have a right to participate and have that participation taken into account in environmental decision-making processes (the principle of access to decision-making); and 3) the public should ultimately have access to an independent and impartial review process, capable of binding public authorities, to allege their rights have been infringed (the principle of access to justice). The Convention is the first time that states have agreed on the minimum content of these principles and established their minimum procedural elements in a single, legally binding international agreement.

APPENDIX 4
THE RESOLUTIONS OF THE PAN EUROPEAN PROCESS

The resolutions adopted by the countries of Europe and the European Union at the conferences held in Strasbourg (1990), Helsinki (1993) and Lisbon (1998) are:

S1:   European network of permanent sample plots for the monitoring of ecosystems.

S2:   Conservation of forest genetic resources.

S3:   Decentralised European data bank on forest fires.

S4:   Adapting the management of mountainous forests to new environmental conditions.

S5:   Expansion of the EUROSILVA Network of research on tree physiology.

S6:   European network for research into forest ecosystems.

H1:   General guidelines for the sustainable management of forests in Europe.

H2:   General guidelines for the conservation of the biodiversity of European forests.

H3:   Forestry co-operation with countries with economies in transition.

H4:   Strategies for a process of long-term adaptation of forests in Europe to climate change.

L1:   People, Forests and Forestry: enhancement of the socio-economic aspects of sustainable forest management.

L2:   Pan-European criteria, indicators and operational guidelines for sustainable forest management.

EC Legislation Concerned with Forest Conservation and Sustainable Development[1]

Development and Utilisation of Woodlands in Rural Areas – Council Regulation (EEC) No.1610/89

This includes a range of measures to promote forest related activities in the context of rural and regional development, including creation and improvement of nurseries; afforestation; improvement of woodlands; protection of forests; development of forest infrastructure; transformation of forestry products; start-up aid for forest management associations

Forestry Measures with Agriculture – Council Regulation (EEC) No. 2080/92

An aid scheme for forestry in agriculture, including afforestation of agricultural land related to the 1992 CAP reform. An EC fund co-finances 50% of the costs. Between 1993 and 1997 some 550,000 hectares of agricultural land were afforested, and 26,000 farmers have taken advantage of the scheme to improve existing woodlands through new investments. But there are concerns that much of this fund has gone to large landowners to the detriment of small farmers, and that the resolution fails to take into account the interests of non-farming community members who also claim a right to determine the future of rural areas.[2]

Protected Forests from Atmospheric Pollution – Council Regulation (EEC) No.3528/86

This was established over a decade ago but has been reinforced by subsequent work. By 1996 over 450 projects had been established, mainly concerned with monitoring air pollution effects on forest ecosystems. The most recent EC/UN-ECE report on forest conditions shows forest damage is worsening slightly.

Protecting Forests from Fires – Council Regulation (EEC) No. 2158/92

Forestry Information and Communication System (EFICS) – Council Regulation (EEC) No. 1100/98

Improving Processing and Marketing Conditions for Forest Products

This provides support for forestry operations upstream of industrial processing, including felling, stripping, cutting up, storage, protective treatment and drying.

Notes

[1] Adapted from EC (1998): *Forest Strategy for the European Union.* Also, EC (1998): *First Report on the Implementation of the Convention on Biological Diversity by the European Community.* Directorate-General Environment, Nuclear Safety and Civil Protection.

[2] Elands, B. (2000): "The Disputed Nature of rural Development: Its Implications for Forestry", in MCPFE (2000): *The Role of Forests and Forestry in Rural Development – Implications for Forest.*